Introduction to Scilab

For Engineers and Scientists

Sandeep Nagar

Apress®

Introduction to Scilab: For Engineers and Scientists

Sandeep Nagar
New York, USA

ISBN-13 (pbk): 978-1-4842-3191-3   ISBN-13 (electronic): 978-1-4842-3192-0
https://doi.org/10.1007/978-1-4842-3192-0

Library of Congress Control Number: 2017959878

Cover image by Freepik (www.freepik.com)

Managing Director: Welmoed Spahr
Editorial Director: Todd Green
Acquisitions Editor: Steve Anglin
Development Editor: Matthew Moodie
Technical Reviewer: Massimo Nardone
Coordinating Editor: Mark Powers
Copy Editor: Ann Dickson

Distributed to the book trade worldwide by Springer Science+Business Media New York, 233 Spring Street, 6th Floor, New York, NY 10013. Phone 1-800-SPRINGER, fax (201) 348-4505, e-mail orders-ny@springer-sbm.com, or visit www.springeronline.com. Apress Media, LLC is a California LLC and the sole member (owner) is Springer Science + Business Media Finance Inc (SSBM Finance Inc). SSBM Finance Inc is a **Delaware** corporation.

For information on translations, please e-mail rights@apress.com, or visit http://www.apress.com/rights-permissions.

Apress titles may be purchased in bulk for academic, corporate, or promotional use. eBook versions and licenses are also available for most titles. For more information, reference our Print and eBook Bulk Sales web page at http://www.apress.com/bulk-sales.

Any source code or other supplementary material referenced by the author in this book is available to readers on GitHub via the book's product page, located at www.apress.com/9781484231913. For more detailed information, please visit http://www.apress.com/source-code.

Printed on acid-free paper

Dedicated to my wife, Rashmi, and my daughter, Aliya

Table of Contents

About the Author

Sandeep Nagar, PhD (Material Science, KTH Royal Institute of Technology, Sweden), teaches and consults on the use of Scilab and other open source software for data science and analysis. In addition to teaching at universities, he frequently gives workshops on open source software and is interested in developing hardware for scientific experiments.

About the Technical Reviewer

 **Massimo Nardone** has more than 22 years of experiences in security, web and mobile development, cloud computing, and IT architecture. His true IT passions are security and Android.

He has been programming and teaching how to program with Android, Perl, PHP, Java, VB, Python, C/C++, and MySQL for more than 20 years.

He holds a Master of Science degree in computing science from the University of Salerno, Italy.

He has worked as a project manager, software engineer, research engineer, chief security architect, information security manager, PCI/SCADA auditor, and senior lead IT security/cloud/SCADA architect for many years.

Massimo's technical skills include security, Android, cloud, Java, MySQL, Drupal, Cobol, Perl, web and mobile development, MongoDB, D3, Joomla, Couchbase, C/C++, WebGL, Python, Pro Rails, Django CMS, Jekyll, and Scratch.

He currently works as the chief information security officer (CISO) for Cargotec Oyj.

He has worked as a visiting lecturer and supervisor for exercises at the Networking Laboratory of the Helsinki University of Technology (Aalto University). He holds four international patents (PKI, SIP, SAML, and Proxy areas).

Massimo has reviewed more than 40 IT books for different publishing companies and he is the co-author of *Pro Android Games* (Apress, 2015).

Acknowledgments

I wish to thank Steve, Mark, and the whole team at Apress for bringing out this book in such a nice format. I also wish to thank the Scilab community for answering questions on forums, which helped me learn difficult concepts with ease.

CHAPTER 1

Introduction to Scilab

1.1 Introduction to Numerical Computing

Numerical computing is a field of mathematics where problems are
made to be solved on a computing device. A variety of software tools
exists for this purpose. In fact, MATLAB, Mathematica, Octave, and
Scilab are specialized softwares for this purpose. A developer can also
use general-purpose programming languages such as C, C++, Python,
and Julia to define mathematical problems, but specialized softwares
present predefined libraries that have been either optimized for stability or
speed or, in same cases, for both. They can also be produced on targeted
hardware for a chosen hardware platform. Thus, using specialized software
is a good way to learn the numerical recipes for physical problems in
an easy fashion. But, the primary question is why we should go for a
numerical solution at all.

The postwar era has seen tremendous advances in computing
technologies as well as in associated software development for specific
purposes. Cost-effective computing resources have made it possible to
simulate almost all physical problems. As a result, numerical computing
has developed as a separate area of mathematics that constantly borrows
and contributes to other areas of mathematics, creating an ecosystem of
computing resources for model-based simulation of physical reality.

© Sandeep Nagar 2017
S. Nagar, *Introduction to Scilab*, https://doi.org/10.1007/978-1-4842-3192-0_1

The need for numerical computing has its roots in the difficulty of deriving analytical solutions for problems. Analytical solutions start with a super-simplified version of physical reality. Such solution are then refined by adding complexities and defining additional parameters to the governing equation(s). But, most of the time, the efforts to just define the problem grow exponentially to the point that it becomes difficult even when you have a team of humans, let alone one person. Thus, the need for an alternative way became apparent, which came in the form of numerical computing.

While analytical computation requires only a pencil, paper, and the human mind, numerical computation requires a calculating device, or a computer. Successful implementation of a computing device to solve problems (especially involving repeated tasks) over a large array of data points was first observed in many fields of science and engineering, for example, when breaking an enemy's secret codes and simulating nuclear reactions before nuclear explosions. The scope of numerical computation was further expanded for civilian purposes. Designing and simulating waterways, dams, electric power stations, and urban roadways are just a few areas of engineering where similar techniques were utilized. All of these applications needed to solve an equation or system of equations for a physical model representing a physical problem.

There are two ways in which we can solve these equations—namely, by using analytical methods and by implementing numerical techniques. We will only concentrate on numerical methods of solving equations in this book.

As time progressed, various schemes to define analytical functions— like differentiation, integration, and trigonometric—were written for digital computers. This involved their digitization, which certainly introduced some errors. Knowledge of the introduced errors and their proper nullification yielded valuable information quicker than analytical results. Thus, it became one of the most actively researched fields of science and continues to be one to this day. The search for faster and accurate algorithms continues to drive innovation in the field of numerical computing and enables humanity to simulate otherwise impossible tasks.

1.2 Various Software Alternatives

A number of alternatives exists to perform numerical computation. Programming languages written to handle mathematical functions, such as FORTRAN, C, Python, and Java, can be used to write algorithms for numerical computation. A set of specialized softwares exists that include MATLAB, Scilab, and Mathematica. Their rich libraries now run in many GBs of data. MATLAB has been tremendously popular among the scientific community since 1984. The cheap availability of digital computing resources propelled its usage in industry and academia to such an extent that virtually every lab needed MATLAB. Even though it wasn't very expensive for a relatively rich western world, it proved to be a costly piece of software for the rest of the world, particularly third-world countries. This part of the world, which has a rich pool of scientists, needed an open source alternative to MATLAB. The solution came in the form of Octave and Scilab. Scilab is extremely powerful, yet it was not designed exactly along the lines of MATLAB, syntax-wise. On the other hand, Octave was developed so that MATLAB's .m files could directly run on Octave. Scilab is now being developed for developers who would like to shift their work to an open source platform completely but do not mind editing their MATLAB files if required.

When compared to the MATLAB package, including specialized packages for various engineering domains and Simulink for the graphical programming paradigm, Octave presents a limited variety of solutions. It includes almost all the functions found in MATLAB's basic package and some specialized packages, but it does not provide Simulink's equivalent. One the other hand, Scilab proves to be a complete alternative since Scilab's XCOS provides Simulink's equivalent, and specialized packages of Scilab can rival the packages provided by MATLAB's functionalities. In this book, we will introduce their basic usage.

Other alternatives include programming languages such as Python, C, C++, and Java. They each have advantages as well as disadvantages, so we advise users to consider their own needs when they make their choice. Scilab is a good choice for prototyping a problem quickly and checking the results. General-purpose programming languages turn out to be a better alternative while working with web-based data collection, analysis and visualization, and implementation of a physical problem on a physical computing platform like FPGA boards, microcontrollers, and so on. Scilab is a high-level language, primarily intended for numerical computation, and it has a rich library of tools for solving numerical linear algebra problems, finding the roots of nonlinear equations, integrating ordinary functions, manipulating polynomials, and solving ordinary as well as partial-differential and differential algebraic equations. This makes it suitable for most basic numerical computational work.

1.3 History

The history of Scilab's creation starts in the 1980s [1]. It was inspired by MATLAB's Fortran code, which was developed by Cleve Moler. Moler later cofounded the MathWorks company with John Little. Scilab initially started as Blaise, a CACSD (computer-aided control system design) software. It was created at the IRIA (French Institute for Research in Computer Science and Control) by François Delebecque and Serge Steer. In 1984, Blaise became Basile and was distributed for a few years by Simulog, the first Inria (French National Institute for Research in Computer Science and Control) startup.

In the 1990s, Simulog stopped distributing Basile. Inria renamed the software Scilab and continued its developed with just six researchers. Scilab 1.1 was released on January 2, 1994, under an open source license. Until version 2.7, the Scilab community grew quite large. With support from companies and institutions, the Scilab Consortium was created in 2003 by Inria. In June 2017, Scilab Enterprises was formed and took charge of development. Based on the classic open source business model, Scilab

Enterprises also offered professional services and support for Scilab. Scilab Enterprises joined the ESI group in 2017 and now develops as well as provides services related to Scilab. Scilab is in use in every strategic domain of science, industry, and services including space, aeronautics, automobile, energy, defense, finance, and transport.

1.4 Installation

Scilab is an open source software distributed under CeCILL license [2], which is a version of GNU-GPL compatible licenses for free software. Please note that the following instructions have been tested on version 6.0.0. It can be downloaded from its web site [3] as per the requirements of the operating system. Installation is quite straightforward and user forums and simple Google searches yield answers to common problems encountered by users. Scilab includes a GUI-based console for easy user interface.

The online implementation of software is available at an Indian web site [4]. In this case, users do not need to install the software on their own systems. They simply need to connect to a remote server using a web browser and they can run the Scilab program in a manner similar to local installation. This does, of course, require a rather fast Internet connection. (Users are encouraged to test this facility.) For the purpose of this book, we will recommend a local installation and work on the same.

After proper installation at a windows OS, users can see a Scilab icon among the list of installed programs. Double-clicking the icon will open a Scilab session. The same is true for MacOSX. For Linux-based OS distributions, you can choose either to click a desktop icon or simply to type `scilab` at the command line.

1.5 Workspace

Figure 1-1 presents a screenshot of installed Scilab version 6.0.0. You can observe various partitions in the main windows:

- File browser

- SCILAB 6.0.0 console

- Variable browser

- Command history

- News feed

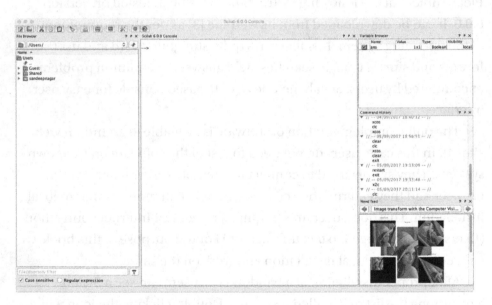

Figure 1-1. *Scilab screenshot*

All these partitions have options of undocking (separating from the main window and opening it in another window) and closing as per requirements. We recommend keeping all these partitions open for now so first-time users can observe the computation and their results in an interactive way.

To check if installation has been done for the version selected for download, we can type a simple command ver at the Scilab command prompt (shown as --> at the Scilab console window). The following output is displayed at the time of writing:

```
1    -->ver
2    ans =
3
4
5    column 1
6
7    ! Scilab Version:                    !
8    !                                    !
9    ! Operating System:                  !
10   !                                    !
11   ! Java version:                      !
12   !                                    !
13   ! Java runtime information:          !
14   !                                    !
15   ! Java Virtual Machine information: !
16   !                                    !
17   ! Vendor specification:              !
18
19   column 2
20
21   ! 6.0.0.1487071837
           !
22   !
           !
23   ! Mac OS X 10.12.6
           !
24   !
           !
```

```
25   ! 1.8.0_51
             !
26   !
             !
27   ! Java(TM) SE Runtime Environment (build 1.8.0_51—b16)
             !
28   !
             !
29   ! Java HotSpot (TM) 64—Bit Server VM (build 25.51—b03,
     mixed mode)
             !
30   !
             !
31   ! Oracle Corporation
             !
```

This output shows that the Scilab version is 6.0.0.1487071837 on MacOSX 10.12.6.

1.6 Command Prompt

Scilab presents a full-featured, interactive, command-line REPL (read-eval-print loop). The interactive shell of the Scilab programming language is commonly known as REPL because it does the following:

- *Reads* what a user types

- *Evaluates* what it reads

- *Prints* out the return value after evaluation

- *Loops* back and does it all over again

This kind of interactive working environment proves very useful for interactive coding as well as for debugging. You can check the results of a particular code as soon as you finish writing it. The way to work with Scilab's REPL is to write the code, analyze the results, and continue this process until the final result is computed. In addition to allowing a quick and easy evaluation of Scilab statements, it also showcases the following:

- A searchable history

- Tab completion

- Many helpful key bindings

- Help and documentation

A section of the left side of a Scilab window shows the command history, which shows all the commands used at the command prompt. Clicking a particular command enables its execution at Scilab's command prompt. This is a quick way to repeat commands. Alternatively, users can use the Up and Down arrow keys on a traditional keyboard of a computer to browse through the commands.

The option of tab completion allows users to just type a few characters for a command and then press the Tab key to obtain either a list of options to choose from if there is more than one option, or it just completes the right syntax of the command. This feature is extremely useful since it avoids syntax errors, which are one of the leading causes of bugs.

The key bindings depend on the operating systems. When you click various items on the menu bar (the top of Scilab's main window), you can see the key bindings alongside the name of the options.

Getting help with various topics and locating documentation is also quite easy in Scilab. You can simply put any argument as a string (characters enclosed within double quotes " ") to the built-in function help(). For example, we used the function ver earlier. Suppose you wish

9

to know its proper usage and wish to see the complete documentation. You can type the following at the Scilab command prompt:

```
1    -->help("ver")
```

This will open up a help browser window, as shown in Figure 1-2.

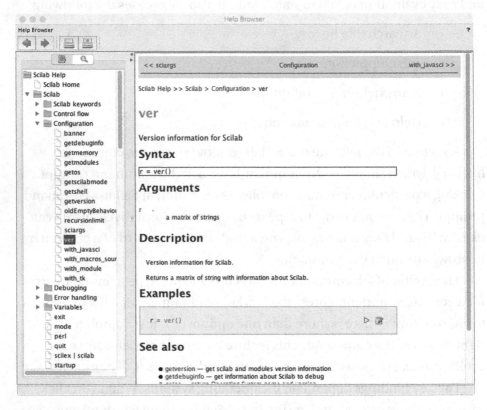

Figure 1-2. *Help browser window*

The help browser hosts all the built-in functions as well as other topics. You can learn Scilab using this browser window and can consult this window in case any questions come up during the development process. The Scilab community maintains very high-quality documentation, which has helped Scilab to become popular.

Another way to get help on a topic is by clicking the ? symbol in the menu bar, which opens up Scilab Help as one of the options. This requires the use of a mouse to locate the symbol and click.

Sometimes users need to obtain a clear screen. To do this, they can use the clc command. This command presents a fresh command prompt, just as in the case of opening Scilab for the first time. It is worth noting that Scilab does not restart during this process. It stores all the variables and their respective values; it just clears the screen o all the previous content. Users can also use this command as a function by typing clc(n) where n can be an integer so that the command will clear those many numbers of lines from the previous session. If users type clc() instead, all lines are cleared from the screen.

The symbol used for this Scilab command is --> to distinguish it from the MATLAB or Octave command prompt. This information is stored in the prompt variable and can be scanned as follows:

```
1    --->prompt
2    ans =
3
4    --->
```

REPL replies with the value --> i. e. symbol for Scilab prompt.

1.7 Variable Browser

The variable browser lists the variables (explained later) used, their sizes, types, and visibility. Having this information on the screen helps users to track the memory usage and also avoid mixing up variable names. The variable browser is appended with new information when users add new variables or remove the same ones from calculations. The variable list can be emptied by the command clear, which removes the variables from the computer memory. This action can be verified by the empty contents of the variable browser.

11

After any kind of activity at the variable browser, there is at least one variable named ans created in a Scilab session that stores the value of the last evaluated expression at the command, as verified by the following code:

```
1    -->2+2
2    ans =
3
4    4.
5
6
7    -->ans
8    ans =
9
10   4.
```

1.8 SciNotes

SciNotes is an embedded Scilab text editor. The SciNotes window can be used to write multiline programs and can be opened via Application -> SciNotes from the menu bar. (See Figure 1-3.)

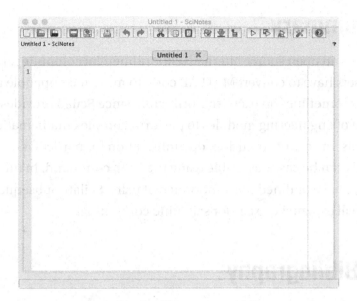

Figure 1-3. *SciNotes window*

It can also be opened using the command prompt by either typing scinotes or editor. The editor has features to create new windows, save the present session, cut, copy, and paste, as well as play the program (in other words, run the program). The output of the code is seen at Scilab's command prompt. The command scinote() makes use of the built-in function. When an argument is given as the filename or filepath, then that particular file opens in the SciNotes environment. The filename must be given as a string. If an additional option, 'readonly', is added, then the file is opened in read-only mode; it cannot be edited. This prohibits unwanted editing of the code.

1.9 Summary

Scilab offers an open source alternative for numerical computing. Even though users have to convert MATLAB code to make it compatible to Scilab, it is sometimes an exercise worthwhile since Scilab provides a rich ecosystem of engineering modules to perform complex mathematical calculations. In case of confusion, documentation for particular commands can be easily available using the help command. In this chapter, we have outlined the importance of using Scilab. Subsequent chapters will explain its usage for scientific computing.

1.10 Bibliography

[1] http://www.scilab.org/scilab/history

[2] http://www.cecill.info/index.en.html

[3] http://www.scilab.org/

[4] http://scilab.in/

CHAPTER 2

Working with Scilab

2.1 Working with Scilab

As explained in Chapter 1, there are two ways to work within Scilab. The first way is to work at the command line by typing one command at a time. The partition titled "Scilab 6.0.0 Console" (shown in the middle part of Figure 1-1) presents the command line where the cursor is shaped as -->. You can write commands here for immediate execution. As an example, let's start with writing the first program of any programming language— printing the words Hello World at the command prompt.

© Sandeep Nagar 2017
S. Nagar, *Introduction to Scilab*, https://doi.org/10.1007/978-1-4842-3192-0_2

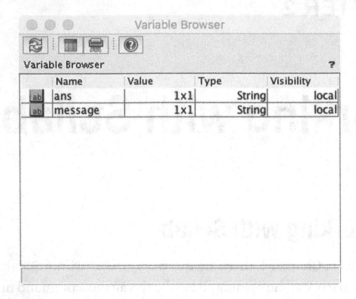

Figure 2-1. *Variable browser window showing two variables*

This can be done by simply writing the following at the command prompt:

```
1  -->message = "Hello World!"
2  message =
3  Hello World!
4  -->disp(message)
5  Hello World!
```

As soon as the first command message = "Hello World!" is given, you can see that the window titled Variable Browser is populated by two variables: ans and message.

The window also states properties of values stored in these variables. Both of them store the same value, which is a string of characters. The ans variable will get appended with new values depending on what is typed next at the command prompt, but the variable message will remain until it is assigned another value.

2.2 Working with Scilab Files

The second method is to write a script (in other words, a set of multiple commands) as a file. Scilab provides two kinds of script files. Two formats, .sci and .sce, differentiate them explicitly:

- .sci files contain Scilab and/or user-defined functions.

  - The execution of such files simply loads them in the Scilab environment but does not execute them.

- .sce files contain both Scilab functions and executable commands. When these files are called, they are *executed*.

Having two types of files is what differentiates Scilab from MATLAB. The distinction between a function file and a file for execution is not present in MATLAB, where all files are simply denoted by an .m extension.

In Listing 2-1, we will write a Scilab file named hello.sce and run it. The file uses a similar code to what we saw in Section 2.1, where we printed Hello World! at the command prompt.

Listing 2-1. hello.sce

```
1  message = "Hello World"
2  disp(message)
```

To run this file, you can simply click the Play icon on the menu bar in the SciNotes window. Alternatively, you can mention the full path of the file to run it. If you are working in the same directory as that of the file, you simply have to print the name of the file without mentioning its extension within the command exec().

The usual conventions of naming the files are followed. The first letter should not be a special character or a numeric value, and file names are case-sensitive. Keywords should be avoided when naming files. It is also recommended that users should work in a single folder for a particular project. The advantage of this approach is that to run programs, users only need to write the name of the program instead of the full path of the file. Also, it is easier to share the codes for executions because paths will differ with different computers.

2.3 Second Example: A Mathematical Calculation

You can write the first program printing the string "Hello World!" by typing the following at the command prompt:

```
1   -->a = 2
2   a =
3   2.
4   -->b = 3
5   b =
6   3.
7   -->a+b
8   ans =
9   5.
10  -->a-b
11  ans =
12  -1.
```

First, the variable a is assigned to a numeric value, 2. Even though it was defined as an integer, it was treated as the floating point number 2, signifying the numeric value 2.0. Similarly, the variable named b is stored as the numeric value 3.0. These values can then be used to perform calculations using these variables. The word *variable* means that they can store *various values*. At any instant of time, the values that they store will be used for calculations. Hence, a+b is calculated as 2.0 + 3.0 = 5.0, and a-b

18

is evaluated as 2.0 − 3.0 = −1.0. The variable named ans stores the value of the last evaluated Scilab expression. These concepts are demonstrated in Listing 2-2.

Listing 2-2. addsub.sce

```
1  a = 2
2  b = 3
3  summing = a+b
4  subtracting = a−b
5  disp(summing)
6  disp(subtracting)
```

Alternatively, you can store these commands by assigning values, calculating them, and then printing them in a file, namely addsub.sce. You can run this file by typing exec(addsub.sce). You can also run the program by simply clicking the Play button in the menu bar. The results can be seen in Scilab's command prompt and found to match the expectations.

To edit the contents of a file, you need to open the editor (SciNotes). You can either open it using the icon provided at the top-left corner of the menu card, or you can type edit addsub.sce at the command prompt. A new window pops up where you can write the code and save it as addsub.sce or a different name as desired. If you don't define the format of the file as .sce, Scilab gives if the format .sci since it treats the file as a Scilab function.

The Scilab command prompt is represented by the symbol --> by default. Scilab is an interpreted language with dynamically typed objects. This implies that after entering a command at the command prompt, if the Enter key is pressed on a keyboard, the command is executed. You can write the code one line at a time and execute the same. The code executes well until an error is presented.

One way to learn Scilab is to run demonstration programs. Click the ? symbol in the list the menu presents at the top of the main window. Choose *Demonstrations* and click *Introduction: Getting started with Scilab.*

Choose the appropriate topic to learn the same. This approach requires users to structure their learning experience themselves; thus, it is not recommended for the first-time user. This book will help with this aspect since the learning structure has been devised for the first-time user.

2.3.1 Calculator

Because the Scilab command prompt executes any expression at the command prompt, Scilab effectively works as a calculator with mathematical operators including multiplication (symbol is *), division (symbol is /), addition (symbol is +), subtraction (symbol is -), and exponentiation (symbol is ^):

```
1  -->2+4.2
2  ans =
3  6.2
4  -->2+4
5  ans =
6  6.
7  -->2-4.2
8  ans =
9  - 2.2
10  -->2*4.2
11  ans =
12  8.4
13  -->2/4.2
14  ans =
15  0.4761905
16  -->2^4
17  ans =
18  16.
```

This example shows that when a command is fed at the command prompt --->, it is executed and the answer is given by displaying the results in the next line as ans =.

2.4 Formatting Command Prompt Display

The display of numbers at the command prompt can be formatted in two ways: in scientific notations (also known as exponent notation) or in normal notations. This is done by issuing the command format e, which prints all answers in scientific notation. At any point, if the command format v is issued, variable adaptive formatting is chosen. Here Scilab chooses either the direct representation or the exponential/engineering/ scientific notation in order to cope with each processed value, as well as with the required width, with a maximal number of output digits. This behavior is shown in the following code:

```
1   -->format e
2   -->2/4.2
3   ans =
4   4.762D-01
5   -->format v
6   -->2/4.2
7   ans =
8   0.4761905
```

When the format() function is used, it can take two inputs—the mode and width of characters—to display. The width of numbers denotes the number of output characters used. This includes the following:

- Sign of the mantissa

- Its digits

- Decimal separator

- Exponent symbol

21

- Sign

- Digits of the exponent

The default width is 10 and the minimal width in 'e' mode is 8. The following sample code explains this behavior:

```
1  -->format('e',10)
2  -->2/4.2
3  ans =
4  4.762D-01
5  -->format('e',20)
6  -->2/4.2
7  ans =
8  4.7619047619048D-01
9  -->format('v',20)
10 -->2/4.2
11 ans =
12 0.4761904761904762
```

When the --> format('e',20) command is issued, 20 digits can be used to display the result so more digits get represented on the screen. Displaying the result with more digits does not alter the storage of the result. The result is stored with a precision of a fixed number of bits, as we will discuss in Section 2.11.1. merely its display is affected.

2.5 Operator Precedence

When a number of operations needs to be performed, the BODMAS rule is followed. Mathematical operations have a precedence order as follows:

- Any expression in brackets is solved first.

  - If brackets further enclose brackets, the inner brackets are solved first and then the successive outer brackets are solved.

- Division

- Multiplication

- Addition

- Subtraction

As shown in the following code, placing brackets has an immediate effect in calculations as brackets are solved first. Hence, we advise that longer mathematical statements are enclosed in brackets appropriately:

```
1  -->(2^2)/3
2  ans =
3  1.3333333
4  -->2^(2/3)
5  ans =
6  1.5874011
```

2.6 Variable Browser Window

As soon as the answer is displayed, the variable browser window shows that a new variable named ans has been created and the value of calculations is stored by this variable name. ans is the default variable name for Scilab. During the course of programming, all variables and their values as well as their type are displayed at the variable browser window.

Double clicking a variable name or any of its properties opens the same as in a graphic interface, which looks similar to an Excel sheet. This is because Scilab performs matrix calculations and, hence, stores values as matrices. A single value is a 1×1 matrix (a matrix having one row and one column). When users learn to define arrays, they can correlate the fact that variables will store values as an $n \times m$ dimensional matrix where data is stored in n rows and m columns. Apart from these two-dimensional arrays, users can also define multidimensional arrays.

2.7 Clearing Variables

Variable values are stored at specific memory locations. The command clear kills these variables, unless they are protected. Protected variables are standard libraries and variables with the percent prefix. As soon as you type clear at the command prompt, you can see that the variable browser is cleared of all values and information.

If three variables a, b, and c have been defined so that they can be selectively killed by issuing the command clear a where only the variable named a is killed, the other two safe. The function isdef() checks whether a variable is live. Listing 2-3 illustrates this point.

Listing 2-3. clear.sce

```
1   disp("Define three vaiables named a,b,c")
2   a = 10;
3   b = 12.2;
4   c = a*b;
5   disp("Checking if variables have been defined as a,b,c")
6   disp(isdef("a"))
7   disp(isdef("b"))
8   disp(isdef("b"))
9   disp("After clearing the variable a")
10  clear a
11  disp(isdef("a"))
12  disp(isdef("b"))
13  disp(isdef("b"))
```

When executed (by typing exec(clear.sce)), it produces the following output:

```
1   -->exec('clear.sce')
2   -->disp("Define three vaiables named a,b,c")
3   Define three vaiables named a,b,c
```

```
 4  --->a = 10;
 5  --->b = 12.2;
 6  --->c = a*b;
 7  --->disp("Checking if variables have been defined as a,b,c")
 8  Checking if variables have been defined as a,b,c
 9  --->disp(isdef("a"))
10  T
11  --->disp(isdef("b"))
12  T
13  --->disp(isdef("b"))
14  T
15  --->disp("After clearing the variable a")
16  After clearing the variable a
17  --->clear a
18  --->disp(isdef("a"))
19  F
20  --->disp(isdef("b"))
21  T
22  --->disp(isdef("b"))
23  T
```

2.8 Comments

All programming languages are designed to encourage programmers
to put comments at desired places in the program so that they can be
understood by humans. Scilab uses the double forward slash (//) for this
purpose. Anything written in front of a comment is not executed by the
interpreter. Comments are used as texts explaining the flow of information
and for tagging information at specific locations in the program.

It is a good practice to write comments for each relevant line of code for better understanding. This is particularly important if more than one person shares the code. The end user can understand the usage of variables and the logic behind calculation by reading the comments.

2.9 Predefined Constants

Scientific computing requires some predefined constants that are frequently used. They are preceded by the % sign, as shown in Listing 2-4.

Listing 2-4. constants.sce

```
1  disp("value of pi:")
2  disp(%pi)
3  disp("value of eps:")
4  disp(%eps)
5  disp("Value of infinity:")
6  disp(%inf)
7  disp("Value of e:")
8  disp(%e)
9  disp("Value of imaginary number i:")
10  disp(%i)
11  disp("value of boolean True:")
12  disp(%t)
13  disp("Value of boolean False:")
14  disp(%f)
15  disp("value of Not-a-number variable i.e. nan:")
16  disp(%nan)
```

When executed (by typing exec(`constants.sce`)), it produces the following output:

```
1   value of pi:
2   3.1415927
3   value of eps:
4   2.220D−16
5   Value of infinity:
6   Inf
7   Value of e:
8   2.7182818
9   Value of imaginary number i:
10  i
11  value of boolean True:
12  T
13  Value of boolean False:
14  F
15  value of Not−a−number variable i.e. nan:
16  Nan
```

A number of physical constants are predefined:

- pi
- e (Euler's number)
- i is the imaginary number $= \sqrt{-1}$
- inf (infinity $= \infty$),
- NaN (Not a Number—resulting from undefined operations, such as Inf/Inf) and eps, which is defined as 2^{-52}

2.9.1 Common Mathematical Functions

A number of predefined mathematical functions exists in Scilab. Following are some of them:

- **Absolute value**: abs() gives the positive value of argument

- **Logarithm**: Natural logarithm log(), Base 10 logarithm log10()

- **Trigonometric functions**: sin(), cos(), tan()

  (Arguments are taken in radians.)

- **Inverse-trigonometric functions**: asin(), acos(), atan()

```
1  -->abs(%i) // absolute value of sqrt(-1)
2  ans =
3  1.
4  -->abs(%pi) // absolute value of pi
5  ans =
6  3.1415927
7  -->abs(-%pi)
8  ans =
9  3.1415927
10 -->log(10) // logarithmic valaue to the base e
11 ans =
12 2.3025851
13 -->log(-10)
14 ans =
15 2.3025851 + 3.1415927i
16 -->log(%e)
```

```
17   ans =
18   1.
19   -->log(-%e)
20   ans =
21   1. + 3.1415927i
22   -->log10(10) // logarithmic valaue to the base 10
23   ans =
24   1.
25   -->log10(-10)
26   ans =
27   1. + 1.3643764i
28   -->sin(45) // sine value
29   ans =
30   0.8509035
31   -->sin((45*%pi)/180) // argument converted to degrees
32   ans =
33   0.7071068
34   -->sqrt(2) // square root
35   ans =
36   1.4142136
37   -->acos(1) // inverse cosine
38   ans =
39   0.
```

Complex calculations using these functions and operations can be performed with ease, as demonstrated in Listing 2-5.

$$\sqrt{sin(10)^2 + cos(10)^2}$$

Listing 2-5. complexCal.sce

```
1  theta = 10; // Theta is taken as 10 here
2  s = sin(theta);
3  s_sq = s^2;
4  c = cos(theta);
5  c_sq = c^2;
6  result = sqrt(s_sq + c_sq); // Variable named "result"
   stores value of answer
7  disp(result)
```

The result is displayed as 1 in compliance with trigonometric identity:

$$\sqrt{sin(\theta)^2 + cos(\theta)^2} = 1$$

2.10 Variable

Variables are used to store values temporarily at computer memory locations. They are actually references to memory locations that store the data. They are addressed using a alphanumeric symbol or set of symbols called *strings*. The example of this approach is given in Scilab code `complexCalc.sce` (see Section 2.9.1) where the numeric value 10 is stored in the variable named `theta`. Then $sin(\theta)$ is calculated and stored in the variable named `s`. Similarly $cos(\theta)$ is stored as `c`. Their respective squares are stored in variables named `s_sq` and `c_sq`. Finally, a variable named `result` stores the calculation $\sqrt{sin(\theta)^2 + cos(\theta)^2}$, and this variable name is used to print the final answer of the calculation. The main advantage of using this approach is that value allocation to a variable is a dynamic process. If a user changes the value of the variable `theta` to 5 instead of 10, the value changes at all places where this variable is called.

2.10.1 Assignment Operator =

The symbol "=" works as an assignment operator that assigns the value present on the right-hand side to the variable name on the left-hand side. In mathematics, the symbol is used to equate two sides of an equation, but in Scilab, the symbol is used for assigning values. The equivalent of a mathematical symbol for equality is the symbol ==, which checks the equality of data value for quantities on both sides:

```
1    --->1==2
2    ans =
3    F
4    --->1==1.0
5    ans =
6    T
7    --->a=1
8    a =
9    1.
10   --->b=2
11   b =
12   2.
13   --->a==b
14   ans =
15   F
```

In this case, $1 \neq 2$. Hence, the result is the boolean value F, signifying false. The numerical values of 1 and 1.0 are the same, so when their equality is probed, you obtain T, signifying the expression is evaluated to be True. Variables can also be used for evaluation. In this case, the latest value stored in a variable is used for evaluation, and the result is declared as a boolean value.

Multiple assignments can be performed by using the comma (,) operator. Also, if we do not wish to produce the results on screen, we can suppress this by using the ; operator:

```
1   -->a1=10, a2=20, a3=30
2   a1 =
3   10.
4   a2 =
5   20.
6   a3 =
7   30.
```

Printing output can be suppressed using the ; operator in front of the command. In the following case, the assignment for a3 is not produced on the terminal since it is suppressed with the ; operator. In the second case, all outputs are suppressed:

```
1    -->a1=10, a2=20, a3=30;
2    a1 =
3    10.
4    a2 =
5    20.
6    -->a3
7    a3 =
8    30.
9    -->a1=10; a2=20; a3=30;
10   -->
```

Note that the only printing of the output at the terminal is suppressed, but the assignment operation is unaffected by the ; operator. It can still be accessed as shown by the second command at the terminal.

2.10.2 Naming Conventions for Variables

There are some naming conventions for variables names that must be
respected to avoid errors:

- Names should not start with a number; however,
  numbers can be used anywhere afterward.

- Variable names are case-sensitive.

- Keywords cannot be used as names.

- Names can include underscore (_) but cannot include
  a whitespace.

2.10.3 Global and Local Variables

A variable declared globally within the main program is known as a *global
variable*, whereas a variable declared locally within a function is known
as *local variable*. Global variables are available to all functions, while local
variables are available only for particular functions. They are defined
using a global declaration statement. Once defined, they remain the same
irrespective of any new definition unless the clear command is issued for
clearing variable names and values from the memory. Another command,
clearglobal, performs the task of killing global variables selectively. To
check if a variable is global, users can use the isglobal() function.

Global variables are used to define constants during numerical
calculations. Suppose we wish that all variables except a select few should
change values. We then name those unchanging values to be global
variables by giving the name of our choice. The predefined variables such
as %pi and %e have been defined in a similar manner.

2.10.4 List of Variables

A list of all variables can be obtained by the commands who and whos where who simply presents the list of variables in the workspace and whos presents the same with more details such as the size of the variable, the number of bytes used to store the variable, and the variable type. who('local') or who('get') returns current variable names and the memory used in double precision words. who('global') returns global variable names and the memory used in double precision words. who('sorted') displays all variables in alphabetical order. By using who and whos, you can keep track of memory requirements judiciously. This is critical in memory- and speed-starved systems like Raspberry Pi.

2.11 Data Types

While assigning data to a variable, it is important to understand that data can be defined as a variety of an object defined by its data type. The command help(type) gives detailed documentation about various data types.

2.11.1 Numerical Data

Numbers are the basic building blocks of numerical analysis. The representation of numbers as computable quantities for a computer requires them to be stored as data in the computer's memory and represent them by a predefined system on various output terminals. Since the memory is limited in nature, fixed spaces of memory are assigned for various number types like integers, real numbers, and complex numbers. In the same manner, various output terminals have fixed capabilities to represent numbers. For example, if the fraction $-\dfrac{2}{3}$ must be displayed, then Scilab must be able to print the numerator and denominator separated by a horizontal line matching the size of both numerals, and the negative sign should be placed at the same height as the horizontal bar, but separated by a distance to distinguish it clearly. Similarly, showing the number as 2×10^{-3} includes storing information about the numbers 2, 10, and −3 separately and displaying them accordingly where −3 must be superscript of 10, and so on. The graphic capabilities of Scilab are quite limited and this topic is out of the scope of this book. Consequently, we will concentrate on the topics related to the storage of numerals.

In the era of cheap computer memory, why should we care to seek less of it for the data? If we have a machine with 64-bit architecture, then it can assign 64 bits for each entity. But would it be wise to use 64 bits to store the small values (say 0)? Automatic assignment faces this inefficient way of computation. Hence, it remains a developer's choice to either declare the data type strictly or let Scilab take care of it. When used judiciously, this quality speeds up computation and lessens the requirements of memory space.

Table 2-1 shows how signed and unsigned integers are stored.

Table 2-1. *Number Datatypes for Integers with Varying Precision*

Data type	Meaning	Range
int8	Signed integer of 8 bits	$\pm 2^8 - 1$
uint8	Unsigned integer of 8 bits	$[0, 2^8]$
int16	Signed integer of 16 bits	$\pm 2^{15} - 1$
uint16	Unsigned integer of 16 bits	$[0, 2^{16}]$
int32	Signed integer of 32 bits	$\pm 2^{31} - 1$
uint32	Unsigned integer of 32 bits	$[0, 2^{32}]$
int64	Signed integer of 64 bits	$\pm 2^{63} - 1$
uint64	Unsigned integer of 64 bits	$[0, 2^{64}]$

Let's test these limits starting with the int8 and unit8 data types. They can be used to convert an input number into an object of the their type by using functions int8() and uint8(), respectively:

```
1   -->2^8
2   ans  =
3   256.
4   -->int8(2^8)
5   ans  =
6   0
7   -->uint8(2^8)
8   ans  =
9   0
10  -->uint8(2^8-1)
11  ans  =
12  255
13  -->int8(2^8-1)
14  ans  =
15  -1
16  -->uint8(2^8+1)
17  ans  =
18  1
19  -->int8(2^8+1)
20  ans  =
21  1
```

$2^8 = 256$ so, with 8 bits, this is the biggest number you can store. However, since the numbers also include 0, an 8-bit storage system can store a maximum of 0 to 255 as numeral values. When you use int8 representation, then one of the bits is used up to represent the signed bit. This leaves the range of $\pm 2^8 - 1$. For this reason, int8(2^8) does not equal 256. But even unit8(2^8) does not equal 256 because the maximum number that can be stored is 255. Why, then, is 0 is shown as answer?

The system of storage follows a cyclic nature where the next number after the maximum number 255 is 0. This can be verified by the fact that uint8(2^8-1) equals 255. On the other hand, int8(2^8-1) is −1 because int8 data type can also store the negative numbers and, hence, −1. Both uint8(2^8+1) and int8(2^8+1) results in 1 since 1 comes after 0.

For storing numbers beyond the range allowed by a particular precision, you must use data types with more precision (more numbers of bits used to store digits). For example:

```
1  -->a = int8(2^10)
2  a =
3  0
4  -->a = int16(2^10)
5  a =
6  1024
7  -->a = int32(2^10)
8  a =
9  1024
```

Since $2^{10} = 1024$, which is beyond the range of int8 data type, it results in the storage of 0 as its value, whereas it can be stored in int16, int32 data types.

Other data types for numerical values are double and float, which store real numbers. Real numbers are represented as floating point numbers in a computer. The mapping of a real number to a computer's storage system is a formulaic representation (called *floating point representation*). For example, the value of $\frac{1}{3}$ is 0.3333333... . Let's suppose that we have only four significant digits for a particular calculations. If so, the value can be rewritten as 3.3333×10^{-1} where 3.3333 is called the *significand*, 10 is called the *base*, and −1 is called the *exponent*. This is further explained in Section 2.11.2.

While assigning a number to the significand, the information about the number of significant digits is used. The significant figures of a number are digits that carry meaningful contribution to its measurement resolution. In the previous case, we assumed only four significant digits, depending on requirements of calculations/measurements. The term *floating point* refers to the fact that a number's *radix point* (decimal point) can "float"; that is, it can be placed anywhere relative to the significant digits of the number. This position is indicated as the exponent component Thus, the floating-point representation can be thought of as a kind of scientific notation.

2.11.2 How to Store Floating Point Numbers

Computers can store numbers as floating point objects. A floating point object stores a number as follows:

$$\pm d_1 d_2 \ldots d_s \times \beta^e \tag{2.1}$$

where $d_i = 0,1,2\ldots\beta - 1$ but $d_1 \neq 0$ and $m \leq e \leq M$ where $m \in I^-$ and $M \in I^+$.

Following are the three parts of a floating point:

- Sign ($\pm$)

- Significand (Mantisa) ($d_1 d_2 \ldots d_s$)

- Exponent (β)

Each part of a floating point number is stored at different memory locations and occupies a specified number of bits. How many bits are defined to which part? These questions have been answered by IEEE standards known as IEEE754. First, let's look at the concept of precision of a number representation:

1. **Single precision**:

    - Occupies 4 bytes = 32 bits

2. **Double precision**:

    - Occupies 8 bytes = 64 bits

3. **Extended double precision:**

   - Occupies 80 bits

4. **Quadruple precision:**

   - Occupies 16 bytes = 128 bits

On a 64-bit operating system machine having Scilab software, real variables are stored as 64 bit floating point numbers (double precision, hence, the name double for the data type). This format includes the following:

- One bit to store the sign of the number

  - $2^1 = 2$ possible values (0 and 1, which are used to represent sign)

- 52 significant bits

  - 2^{52} possible values

- 11 for exponent

  - 2^{11} possible values

This complies with IEEE 754 standard.

Converting between a higher-precision data type to a lower-precision data type can result in saving the computer's memory and speeding up calculations at the cost of precision. These decisions must be taken by the developer beforehand. You do not always need higher precisions. For example, if you are working with dimensions of a bridge and the numbers are represented in units of meters, you can usually work with a precision of $\dfrac{1}{10}^{th}$ m for most measurements involved in civil engineering. However, if you are working with nano materials or talking about atoms when calculations involve precision around Å= $10^{-9}m$, you obviously need better precision. It is important to perform back-of-the-envelop calculations for a particular problem to get an idea about the maximum and minimum numbers expected during the running of a program. Accordingly, you can assign data types. But what if you don't set a data type? In this case, default data types are used like floating point numbers on a 64-bit machine are stored with double-precision floating point numbers by default.

Specialized packages exist for Scilab where arbitrary precision can be used for more accurate mathematical calculations. Even though a detailed discussion of this topic is beyond the scope of this book, users are advised to check out the packages Xnum (maximum precision of 200 digits) [1] and Mupat [2]. Precision plays an important role in producing meaningfully reliable numerical values for critical applications so data types in such cases must be used judiciously.

2.11.3 Formatted Display of Numbers

Controlling the format of displaying the number of digits of a numerical value is one of the most significant abilities of an advanced programming environment. Scilab provides the command format. Its description can be obtained using the command help format. The command format() takes integer values as arguments for displaying the number of digits in a numerical value. The range for this input argument is [2, 25]:

```
1  --->exp(1)
2  ans =
3  2.7182818
4  --->format(5)
5  --->exp(1)
6  ans =
7  2.72
8  --->format(25)
9  --->exp(1)
10  ans =
11  2.718281828459045090790790956
```

$\exp(1)$ represents $e^1 = 2.7182818$, which is also called *Euler's number*. When format(5) is issued as a command, then the answer to e^1 is shown as 2.72, whereas format(255) reveals 2.7182818284590450907956.

It is important to note that the stored value is unaffected by the fact that it is being represented on a graphic terminal with fewer digits. The symbolic representation of a number on a computer terminal is independent of the way it is stored (which is determined entirely by its data type).

2.12 Boolean Data

help(boolean) provides a detailed description about boolean variables namely \%T for true and \%F for false. Boolean variables must be operated using boolean operators such as the following:

- NOT

    – Represented by the symbolic operator ~

    – Negates an input (~%T=%F and ~%F=%T)

- AND

    – Represented by either the symbolic operator & or the function or()

    – Behavior is shown in Table 2-2

    – Bitwise operations are done by the bitand() function

- OR

    – Represented by either the symbolic operator | or the function or()

    – Behavior is shown in Table 2-3

    – Bitwise operations are done by bitor() function

41

- XOR

  - Represented by the function `bitxor()`

  - Behavior is shown in Table 2-4

Table 2-2. *Truth Table for AND Operator*

AND	T	F
T	T	F
F	F	F

Table 2-3. *Truth Table for OR Operator*

OR	T	F
T	T	T
F	T	F

Table 2-4. *Truth Table for XOR Operator*

XOR	T	F
T	F	T
F	T	F

```
1  -->a = %T // a is give value True
2  a =
3  T
4  -->b = ~a // b is NOT a
5  b =
6  F
```

```
7   -->a&b // a AND b
8   ans =
9   F
10  -->a|b // a OR b
11  ans =
12  T
```

The `bitxor()` operator works bitwise, so it converts an input into its binary representation and performs the computation. Let's understand this concept by using an example. `bitxor(1000,1001)` is 1 because binary representations (obtained by the function `dec2bin()`) show that only the last bit is different and, hence, the result is the binary number whose decimal equivalent is 1:

```
1   -->bitxor(1000,1001)
2   ans =
3   1.
4   -->dec2bin([1000;1001;1])
5   ans =
6
7   !1111101000   !
8   !             !
9   !1111101001   !
10  !             !
11  !0000000001   !
12  -->bin2dec("0000000001")
13  ans =
14  1.
```

It is important to note that, in binary logic, the values True and False are equivalently used with 1 and 0. So Tables 2-2, 2-3 and 2-4 would simply replace the value T with 1 and F with 0.

Similarly, the `bitand()` and `bitor()` operators can be understood with the following example:

```
1   -->bitor(100,99)
2   ans =
3   103.
4   -->bitand(100,99)
5   ans =
6   96.
7   -->dec2bin([100;99;103;96])
8   ans =
9
10  !1100100  !
11  !         !
12  !1100011  !
13  !         !
14  !1100111  !
15  !         !
16  !1100000  !
```

2.13 Strings

Alphabets along with special characters are treated as string data types. The built-in function `string()` converts a given value to string data type. As shown in the following example, the numerical value 2 is converted to a string data type when the `string()` function is applied to it and remains an integer type otherwise:

```
1   -->a = string(2)
2   a =
3   2
```

```
4  ——>b = string(3)
5  b =
6  3
7  ——>a+b
8  ans =
9  23
```

It is worth noting that binary data for 2 as a string (stored in variable a) and as a number will be quite different and, thus, the numerical computations will treat them differently even if it is always displayed as 2. When it is represented as a string, operators will first understand the data type and then operate. An addition operator is simply a concatenation operator for strings. Hence, a+b equals 23. A numerical computation of 2 + 3 results in the value 5.

Without the string() function, strings can be defined by enclosing them in single quotes (") or double quotes (""). Strings are case-sensitive, too. This can be verified by the following code:

```
1  ——>a = "hello world"
2  a =
3  hello world
4  ——>b = 'Hello World'
5  b =
6  Hello World
7  ——>a==b
8  ans =
9  F
```

a contains h and w characters (lowercase) while b contains H and W characters. For this reason, a==b gives F (false) since their values are different.

2.14 Complex Numbers

One of the most important parts of doing mathematical calculations on computers is the ability to work with complex numbers and their algebra. Computations involving complex numbers can be found in almost all branches of science and mathematics. Finding flexible ways of defining complex numbers and their mathematics is an art that all developers must employ to compute efficiently.

A graphical description of a complex number is shown in Figure 2-2. On a real-imaginary axis-based complex plane, a particular point is defined by a complex number $a + ib$ where a is the magnitude of the projection of a point on real axis and b is the magnitude of projection of a point on imaginary axis.

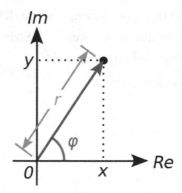

Figure 2-2.   *Complex number depicted on complex plane [3]*

Figure 2-2 shows a point depicting the complex number $z = x + iy$. The value of $r = |z|$ (absolute value) and ϕ (argument) are given by the following:

$$r = \sqrt{x^2 + y^2} \tag{2.2}$$

$$\phi = tan^{-1}\left(\frac{y}{x}\right) \tag{2.3}$$

The absolute value of a complex number is simply its distance from its origin. The argument of a complex number is simply the angle it makes with the horizontal axis in a counterclockwise direction. The absolute value of a complex number can be calculated using the abs() function:

```
1  --->complex(2,3)
2  ans =
3  2. + 3.i
4  --->abs(complex(2,3))
5  ans =
6  3.61
```

The angle can be calculated by using real and imaginary parts of the complex number as follows:

```
1  --->a = complex(2,3)
2  a
3  2. + 3.i
4  --->type(a)
5  ans =
6  1.
7  --->real(a)
8  ans =
9  2.
10 --->imag(a)
11 ans =
12 3.
13 --->atan(imag(a)/real(a))
14 ans =
15 0.98
```

2.14.1 Real and Imaginary Parts

Real and imaginary parts of a real number are given by the built-in functions real() and imag() as follows:

```
1   --->r = real(complex(2,3))
2   r =
3   2.
4   --->i = imag(complex(2,3))
5   i =
6   3.
7   --->r = real(complex([2,3],[-3,-4]))
8   r =
9   2. 3.
10  --->i = imag(complex([2,3],[-3,-4]))
11  i =
12  -3. -4.
```

2.14.2 Complex Conjugates

The complex conjugate of a complex number $x + iy$ is $x - iy$ and that of $x - iy$ is $x + iy$. The built-in function conj() produces the complex conjugate of a given complex number:

```
1   --->a = complex(2,3)
2   a =
3   2. + 3.i
4   --->b = conj(a)
5   b =
6   2. - 3.i
7   --->a*b
8   ans =
9   13.
```

The product of a complex number with its conjugate can be understood as

$$\prod\left((a+ib),(a-ib)\right)=a^2+b^2 \qquad (2.4)$$

where a and b are real and imaginary parts.

For multiple entries (complex numbers) as array elements:

```
1  -->a = complex([1 2 3],[4 5 6])
2  a =
3  1. + 4.i     2. + 5.i     3. + 6.i
4  -->b = conj(a)
5  b =
6  1. - 4.i     2. - 5.i     3. - 6.i
7  -->a.*b
8  ans =
9  17.   29.   45.
```

2.14.3 imult

The built-in function imult() is an efficient way of multiplying a real number with i (the imaginary unitary) especially when %inf and %nan are present in the calculation:

```
1  -->imult(%nan)
2  ans =
3  Nani
4  -->imult(%inf)
5  ans =
6  Infi
7  -->%i*%nan
8  ans =
9  Nan + Nani
```

```
10  -->%i*%inf
11  ans =
12  Nan + Infi
13  -->imult(4)
14  ans =
15  4.i
16  -->imult(complex(4,4))
17  ans =
18  -4. + 4.i
```

As seen when

$$4 \times i = 4i$$

and

$$(4 + 4i) \times i = 4i - 4 = -4 + 4i$$

2.14.4 Checking If a Variable Has Complex Components

The built-in function isreal() checks if a variable has complex components. It produces an boolean output T if the input variable does not have complex numbers and outputs F otherwise:

```
1  -->a = [1 2 3]
2  a =
3  1.    2.    3.
4  -->a1 = complex(a,a)
5  a1 =
6  1. + i       2. + 2.i       3. + 3.i
7  -->isreal(a)
8  ans =
9  T
```

```
10  -->isreal(a1)
11  ans =
12  F
```

2.14.5 Complex Arithmetic

Complex arithmetic involves similar operations as for real numbers such as addition, subtraction, multiplication, division, raised to a power, and so on. Rules for complex numbers are a bit different for these operations.

Adding two complex numbers involves adding their real and imaginary parts. So is the case with subtraction. Suppose we define two complex numbers as follows:

$$z_1 = a_1 + b_1 i$$
$$z_2 = a_2 + b_2 i$$

Then you can define their addition and subtraction:

$$z_1 + z_2 = (a_1 + a_2) + (b_1 + b_2)i$$
$$z_1 - z_2 = (a_1 + a_2) - (b_1 + b_2)i$$

Multiplication and division operations for complex numbers are not so straightforward:

$$z_1 \times z_2 = (a_1 \times a_2) + (a_1 \times b_2)i + (a_2 \times b_2) + (b_1 \times b_2)(i^2)$$

which simplifies by collecting real terms and imaginary terms:

$$z_1 \times z_2 = (a_1 a_2 - b_1 b_2) + (a_1 b_2 + a_2 b_1)i$$

because $i^2 = -1$. Multiplying and dividing a complex number with a real number can be done in a simpler manner by performing the multiplication or division for real and imaginary part, respectively.

51

A complex conjugate of a complex number $z_1 = a_1 + b_1 i$ is defined as $z_1^* = a_1 - b_1 i$. Geometrically, z_1^* is the "reflection" of z_1 about the real axis. Hence, if we calculate the conjugate twice, we get the same number: $(z_1^*)^* = z_1$.

Division of a complex number can be performed using its conjugate as follows:

$$\frac{a_1 + b_1 i}{a_2 + b_2 i} = \frac{a_1 + b_1 i}{a_2 + b_2 i} \times \frac{a_2 - b_2 i}{a_2 - b_2 i} = \frac{a_1 a_2 + b_1 b_2}{a_2^2 + b2^2} + \frac{b_1 a_2 - a_1 b_2}{a_2^2 + b2^2} i$$

Thus, multiplying the denominator's complex conjugate with both the numerator and denominator yields a new complex number that is the result of division of two complex numbers. This can be checked for two complex numbers, $z_1 = 2 + 3i$ and $z_2 = -3 + 4i$, as follows:

```
1   -->z1 = complex(2,3)
2   z1 =
3   2. + 3.i
4   -->z2 = complex(-3,4)
5   z2 =
6   -3. + 4.i
7   -->z1+z2 // summing two complex numbers
8   ans =
9   -1. + 7.i
10  -->z1-z2 // subtracting two complex numbers
11  ans =
12  5. - i
13  -->z1*z2 // multiplying two complex numbers
14  ans =
15  -18. - i
```

```
16  -->z1/z2 // dividing two complex numbers
17  ans =
18  0.24-0.68i
19  -->z1^2 // complex number raised to the power 2
20  ans =
21  -5. + 12.i
22  -->z1^z2 // complex number raised to the power another
    complex number
23  ans =
24  -2.0D-04 + 3.0D-04i
25  -->2*z1 // complex number multiplied by a real number
26  ans =
27  4. + 6.i
28  -->z1/2 // complex number divided by a real number
29  ans =
30  1. + 1.5i
31  -->z1+2 // complex number added with a real number
32  ans =
33  4. + 3.i
34  -->z1-2 // complex number and real number subtracted
35  ans =
36  3.i
```

A real number can be understood as a complex number with the imaginary part as zero. When addition and subtraction of a complex number is performed, only the real part is operated. When a complex number is multiplied and/or divided by a real number, each real and imaginary part is operated accordingly.

2.15 Summary

In this chapter, we have discussed the usage of different types of numbers and strings. We have illustrated operating on numbers and strings. It is important to keep in mind that all numerical computations are approximations because, due to the finite precision of numbers expressed on a computing machine, errors are introduced in the computation. Users can choose the precision of a number before a computation, but this should be done cautiously. Working with low-precision numbers occupies less storage and also computes faster; however, it is more inaccurate. Scilab enables defining complex numbers and these defined complex numbers can be operated using the same arithmetic operators as that of real numbers without any modification even though their rules of arithmetic are different.

2.16 Bibliography

[1]   https://www.awitness.org/unifiedm/xnum.html

[2]   https://www.awitness.org/unifiedm/qupat.html

[3]   https://en.wikipedia.org/wiki/File:Complex_
      number_illustration_modarg.svg

CHAPTER 3

Working with Arrays

3.1 Introduction

Matrix formulation of mathematical problems enables faster numerical computations. For a two-dimensional matrix, elements have a unique row and column index through which users can access them. Rows and columns can be attributed to different properties under study. In this way, users can fit data for two properties as a matrix and then use these matrices for numerical calculations. For example, suppose an element of a row is defined as 1 if a compound is a conductor, 2 if it is a semiconductor, and 3 if it is an insulator. Then the row vector (a matrix composed of only one row) $[1\ 0\ 0\ 3\ 2\ 1\ 3\ 0\ 1\ 0\ 3\ 2\ 1]$ has information about 13 compounds. In a numerical calculation involving the conductive nature of a compound, this row vector (a 13×1 matrix) can be utilized where 13 compounds can be simultaneously scanned mathematically.

Scilab objects named `arrays` deal with defining matrices. Using different properties of this object, users can define various kinds of matrices. Built-in functions for matrix operations make it easier for a programmer to deal with large numbers of data by arranging them as a matrix in the desired format and performing array operations.

© Sandeep Nagar 2017
S. Nagar, *Introduction to Scilab*, https://doi.org/10.1007/978-1-4842-3192-0_3

3.2 Arrays and Vectors

Arrays are particularly interesting since they are used to define vectors, tables, and matrices for scientific computing:

- A 1-D (one-dimensional) array acts as a vector or list.

- A 2-D array can be used as a table or matrix.

- 3-D and more-D arrays can represent multidimensional matrices.

In this way, instead of just pointing to a single number, a variable name can point to a sequential set of numbers. The syntax to define arrays dictates that they must be defined within square brackets [ ]:

```
1   --->a = [1 2 3 4 5] // separated by white space
2   a =
3   1.    2.    3.    4.    5.
4   --->b = [1, 2, 3, 4, 5] // separated by commas
5   b =
6   1.    2.    3.    4.    5.
7   --->c = [1.0 2.0 3.0 4.0 5.0] // arrays of floating point
    numbers
8   c =
9   1.    2.    3.    4.    5.
10  --->matrix22 = [1 2; 3 4] // The ; operator defines new row
11  matrix22 =
12  1.    2.
13  3.    4.
14  --->matrix33 = [1 2 3;4 5 6;7 8 9] // arrays with three rows
15  matrix33 =
16  1.    2.    3.
17  4.    5.    6.
18  7.    8.    9.
```

```
19  -->size(matrix22) // size of array i.e number of rows and
    columns
20  ans =
21  2.   2.
22  -->size(matrix33)
23  ans =
24  3.   3.
25  -->matrix32 = [1 2;3 4;5 6]
26  matrix32 =
27  1.   2.
28  3.   4.
29  5.   6.
30  -->size(matrix32)
31  ans =
32  3.   2.
33  -->size(size(matrix33)) // size also outputs an array
34  ans =
35  1.   2.
```

As seen in this example code, an array can be understood as a matrix
consisting of rows and columns. Thus, users can make a desired sized matrix.
For example, matrix22 is a 2 × 2 matrix and matrix33 is a 3 × 3 matrix,
whereas a is a 1 × 5 matrix. The first number when defining the size gives the
number of rows, while the second number gives the number of columns.
The comma (,) operator operates by defining the next element in the same
row, whereas the (;) operator defines the numbers in the next line/row. The
use of the comma operator is optional. If skipped, it must be replaced by a
whitespace. The built-in function size() outputs a 2 × 1 array where the first
number signifies the number of rows and the second number represents the
number of columns.

If the number of elements in each row/column doesn't match, the user will obtain an error message:

```
1   >> right33 = [1, 2, 3; 4, 5, 6; 7, 8, 9]
2   right33 =
3   1   2   3
4   4   5   6
5   7   8   9
6   -->wrong33 = [2, 3; 4, 5, 6; 7, 8, 9]
7   !--error 6
8   Inconsistent row/column dimensions.
9   wrong33 = [1, 2, 3; 4, 5, 6; 8, 9]
10  !--error 6
11  Inconsistent row/column dimensions.
```

Elements of an array can be any data type as defined in Chapter 2, Section 2.11. The built-in function type( ) can be used to determine the data type of stored values. All elements of an array can be set to a particular data type by the following commands:

```
1   -->a = uint8([1 2 3])
2   a   =
3   1   2   3
4   -->b = uint16([1 2 3])
5   b   =
6   1   2   3
7   -->c = uint32([1 2 3])
8   c   =
9   1   2   3
10  -->type(a)
11  ans   =
12  8.
```

```
13  -->type(b)
14  ans  =
15  8.
16  -->type(c)
17  ans  =
18  8.
```

3.3 Operations on Arrays and Vectors

Operating on arrays has two aspects:

- Operating on two or more arrays

- Elementwise operations

3.3.1 Elementwise Operations

All arithmetic operators—such as +,-,*,/,^, and so on—can be used in both cases. When we need to do elementwise operations, a . is placed before the operator so that elementwise operators become .+,.-,.*,./, and .^. This will become more clear in the following example:

```
1  -->a = [1 2 3; 4 5 6; 7 8 9]
2  a  =
3  1.   2.   3.
4  4.   5.   6.
5  7.   8.   9.
6  -->b = a
7  b  =
8  1.   2.   3.
9  4.   5.   6.
10 7.   8.   9.
```

```
11  --->a. / b
12  ans   =
13  1.    1.    1.
14  1.    1.    1.
15  1.    1.    1.
16  --->a^b
17  !--error 43
18  Not implemented in scilab ...
19  at line 61 of function %s_pow called by:
20  a^b
21  --->a.^b // Element wise operation
22  ans =
23
24  1.     4.     27.
25  256.   3.0 D+03   4.0 D+04
26  8.0 D+05   1.0 D+07   3.0 D+08
```

3.3.2 Matrix Multiplication

A matrix of dimensions $a \times b$ can only be multiplied by a matrix of dimensions $b \times c$, which results in a matrix of dimensions $a \times c$. It is performed by multiplying elements of rows by elements of columns to get new elements.

The following code will illustrate this matrix multiplication method:

```
1  --->a = [1, 2; 3, 4; 5, 6] // Defining a 3 X 3 matrix
2  a =
3  1.   2.
4  3.   4.
5  5.   6.
6  --->size(a) // size is found to be 3 X 2
7  ans =
8  3.   2.
```

```
 9  --->a_t = a' // defining transpose
10  a_t =
11  1.    3.    5.
12  2.    4.    6.
13  --->size(a_t) // size of transpose is 2 X 3
14  ans =
15  2.    3.
16  --->mul=a*a_t // multiplying two matrices
17  ans =
18  5.     11.    17.
19  11.    25.    39.
20  17.    39.    61.
21  --->size(mul) // size of multiplication matrix is 3 X 3
22  ans =
23  3.    3.
```

a' gives the transpose of a matrix (rows are made columns and vice versa). This makes a 3 × 2 matrix become a 2 × 3 matrix. When they are multiplied, you get a 3 × 3 matrix.

3.3.3 Inverse of Matrices

Performing division of a matrix involves *matrix inversion*. An inverse matrix is such that its multiplication with the original matrix yields an identity matrix (a matrix with determinant as 1), that is, a matrix with 1 at its diagonal elements and 0 otherwise:

```
1  --->a = [2 5 4;-4 6 -3;4 7 -1] // Defining a 3 X 3 matrix
2  a =
3  2.    5.    4.
4  -4.   6.    -3.
5  4.    7.    -1.
```

```
6  --->inverse = inv(a) // calculating inverse of matrix a
7  inverse =
8  -0.06    -0.13    0.15
9  0.06     0.07     0.04
10 0.2      -0.02    -0.12
11 --->a*inverse // matrix multiplied by inverse is identity
   matrix
12 ans =
13 1.   0.   0.
14 0.   1.   0.
15 0.   0.   1.
```

The function pinv() is used if the input matrix is a nonsquare matrix. In other words, the number of rows and the number of columns are not equal. The pinv() function gives a psuedo-inverse of a matrix such that if X=pinv(a), then $a \times X \times a = a$, $X \times a \times X = X$, and both $a \times X$ and $X \times a$ are a Hermitian matrix:

```
1  --->a = [1,2,3; -5 4 -7]
2  a =
3  1.    2.    3.
4  -5.   4.    -7.
5  --->inverse = pinv(a)
6  inverse =
7  0.       -0.06
8  0.27     0.1
9  0.15     -0.05
10 --->a*inverse
11 ans =
12 1.   0.
13 0.   1.
```

I is known as an identity matrix because all its diagonal elements are 1 and all its nondiagonal elements are zero, which makes its determinant 1.

3.3.4 det()

The determinant of a matrix a is calculated by the command det(a):

```
1  -->a = [1 2 3;4 5 6;7 8 9]
2  a =
3  1.   2.   3.
4  4.   5.   6.
5  7.   8.   9.
6  -->det(a)
7  ans =
8  -9.0 D-16
```

3.3.5 rank()

The rank of a matrix (the number of linearly independent rows or columns) can be determined by the built-in rank() function:

```
1  -->a = eye(3,3)
2  a =
3  1.   0.   0.
4  0.   1.   0.
5  0.   0.   1.
6  -->rank(a) // all rows are linearly independent
7  ans =
8  3.
9  -->a = [1, 2, 3;2, 4, 6;4, 6, 3] // second row in just
   first rwo multiplied by 2
```

```
10  a =
11  1.    2.    3.
12  2.    4.    6.
13  4.    6.    3.
14  -->rank(a)
15  ans =
16  2.
17  -->a = [1, 2, 3;2, 4, 6;3, 6, 9] // second and third row
    are merely first row multiplied by 2 and 3 res pectively
18  a =
19  1.    2.    3.
20  2.    4.    6.
21  3.    6.    9.
22  -->rank(a)
23  ans =
24  1.
```

3.3.6 trace()

The sum of the diagonal elements of a matrix is called the trace of the matrix. This is given by the built-in trace() function as follows:

```
1  -->a = [1, 2, 3;2, 4, 6;3, 6, 9]
2  a =
3  1.    2.    3.
4  2.    4.    6.
5  3.    6.    9.
6  -->trace(a)
7  ans =
8  14.
```

The automatic generation of an identity matrix is done by using the command eye(a,b) where a and b are values of the number of rows and columns:

```
1   -->eye(2,3)
2   ans =
3   1.      0.      0.
4   0.      1.      0.
5   -->a = eye(4,4)
6   a =
7   1.      0.      0.      0.
8   0.      1.      0.      0.
9   0.      0.      1.      0.
10  0.      0.      0.      1.
11  -->det(a)
12  ans =
13  1.
```

3.3.7 Magnitude of a Vector

A mathematical vector can be defined as a 1 × 3 array. The built-in function norm() gives the magnitude of a vector. This is the Euclidean distance from the origin to the vector. So, if a vector is defined as the following:

$$\vec{a} = A\vec{x} + B\vec{y} + C\vec{z}$$

its magnitude can be defined as:

$$N = \sqrt{A^2 + B^2 + C^2}$$

A unit vector for a vector is obtained by dividing the vector by its own norm/magnitude. This operation can be performed with ease in Scilab:

```
1   -->a = [1 2 3]
2   a =
3   1.    2.    3.
4   -->norm(a)
5   ans =
6   3.74166
7   -->a = [1 2 3]/norm(a)
8   a =
9   0.26726    0.53452    0.80178
```

3.3.8 Random Matrix

Using random number generators, a random matrix can be created by the command rand(a,b):

```
1    -->rand(4,5)
2    ans =
3    0.21132      0.66538      0.87822      0.72635      0.23122
4    0.75604      0.62839      0.06837      0.19851      0.21646
5    0.00022      0.84975      0.56085      0.54426      0.88339
6    0.33033      0.68573      0.66236      0.23207      0.65251
7    -->rand(4,5)
8    ans =
9    0.30761      0.36164      0.33217      0.26931      0.04373
10   0.93296      0.29223      0.59351      0.63257      0.48185
11   0.21460      0.56642      0.50153      0.40520      0.26396
12   0.31264      0.48265      0.43686      0.91847      0.41481
13   -->rand(1)
14   ans =
15   0.28065
```

```
16   -->rand(4,1)
17   ans =
18   0.12801
19   0.77831
20   0.21190
21   0.11214
22   -->rand(1,4)
23   ans =
24   0.68569      0.15312      0.69709      0.84155
```

Please note that the numbers generated in the previous example will be different each time, even on the same machine, since they are supposed to be random in nature. By default, they are uniformly distributed over the interval (0, 1). A vector is simply a row vector so it can be generated randomly by the command rand(a). help("rand") gives a detailed description about various other features and arguments of a random number generator:

```
1    -->rand(4,5, 'uniform')
2    ans =
3    0.11384      0.68540      0.38738      0.37601      0.26386
4    0.19983      0.89062      0.92229      0.73409      0.52536
5    0.56187      0.50422      0.94882      0.26158      0.53762
6    0.58962      0.34936      0.34353      0.49935      0.11999
7    -->rand(4,5, 'normal')
8    ans =
9    - 1.28586    - 0.37782    - 0.45753    0.01164      1.74874
10   0.59712      1.14562      0.56232      0.22327      1.86518
11   0.61078      2.57491      - 0.64533    - 1.43445    0.16459
12   - 1.05679    - 0.50056    - 0.36478    1.73638      - 1.03589
```

The 'normal' and 'uniform' functions provide *normal* and *uniform* distributions of random numbers. The normal distribution has a mean at 0 and variance 1. All the random numbers follow Gaussian distribution with their mean at 0 and distributed between −1 and 1. On the other hand, default random number distribution is 'uniform' and these numbers are generated between 0 and 1.

A more refined built-in function named grand() gives a variety of options to choose distributions and set limits for the set of numbers. Simply typing help("grand") provides detailed documentation for its usage.

3.3.9 Indexing

Each element of the matrix is characterized by two numbers, the row number and the column number. These numbers are used to pinpoint an element and operate on that. They are called an *index* of elements. The assignment operator can be used to set the value at a particular index to produce a new matrix with updated value(s). The following example illustrates these concepts:

```
1 ——>a = rand(3,4)
2 a =
3 0.38850    0.04070    0.44219    0.01109
4 0.67899    0.64673    0.48739    0.84484
5 0.37286    0.81418    0.76592    0.89999
6 ——>a(2,1) // element at row 2 and column 1
7 ans =
8 0.67899
9 ——>a(4,2) // element at row 4 and column 2 does not exist
10  Invalid index.
11  ——>a(2,2)=0 // setting element at row 2 column 2 to 0
```

```
12   a =
13   0.38850      0.04070      0.44219      0.01109
14   0.67899      0.          0.48739      0.84484
15   0.37286      0.81418      0.76592      0.89999
```

It is important to note that, unlike some programming languages where indices start from 0, in Scilab indices start from 1 and Scilab does not take negative numbers as indices.

3.3.10 Using Indices to Make New Vectors

In the following example, b is a new vector formed from vector a where successive elements are made up of elements taken from a index vector [2 4 2 4]:

```
1   −−>a = [1 4 1 3 1 4 1 4 2 4 5]
2   a =
3   1.   4.   1.   3.   1.   4.   1.   4.   2.   4.   5.
4   −−>b = a([2 4 2 4])
5   b =
6   4.   3.   4.   3.
```

The same method can be employed for two-dimensional and higher-dimensional matrices:

```
1   −−>a
2   a =
3   0.61618      0.38019      0.76251      0.40089
4   0.63517      0.35260      0.7284       0.55388
5   0.15988      0.91202      0.27883      0.27174
6   −−>a([1, 2], [2, 3])
7   ans =
8   0.38019      0.76251
9   0.35260      0.7284
```

Please note that since the use of the comma operator is optional, vectors and matrices will be defined by simply inserting a whitespace.

3.3.11 Slicing

Matrices can be sliced to desired portions by using indices and the colon : operator. If n:m is provided for slicing, a new matrix is fabricated where elements with index n to index m are placed. Also the ; operator defines the new column while slicing. Some examples will make the usage clear:

```
1   -->a = [1,2,3;4,5,6]
2   a =
3   1.    2.    3.
4   4.    5.    6.
5   -->B = [(1:3);(4:6);(7:9)]
6   B =
7   1.    2.    3.
8   4.    5.    6.
9   7.    8.    9.
10  -->c = [a;B]
11  c =
12  1.    2.    3.
13  4.    5.    6.
14  1.    2.    3.
15  4.    5.    6.
16  7.    8.    9.
17  -->c = [a,B]
18  inconsistent row/column dimensions
```

Here the matrix a has elements 1, 2, 3 in the first row and then a row separator ; defines the next row of elements 4, 5, 6. Similarly, matrix B is defined by rows defined by commands:

- $(1:3)$, which results in $(1, 2, 3)$

- $(4:6)$, which results in $(4, 5, 6)$

- $(7:9)$, which results in $(178, 9)$

Please note the Scilab variable names are case-sensitive, hence a will not be the same as A. Now a new matrix is created named c by vertically concatenating the matrices a and B. Consequently, the resultant matrix c is made of the elements of a stacked on top of the elements of B. The command c=[a,B] yields an error because the dimensions of a and B are not consistent for horizontal concatenation.

Horizontal concatenation can instead be easily performed in the following way in our case:

```
1   -->a = 1:3 // array whose element start from 1 and ends at 3
2   a =
3   1.   2.   3.
4   -->A = [a,a] // New array with old array as its two
    element in a row
5   A =
6   1.   2.   3.   1.   2.   3.
7   -->A = [a;a] // New array with old array as its two
    elements in a column
8   A =
9
10  1.   2.   3.
11  1.   2.   3.
```

It is worth noting that the comma operator (,) separates the elements of a row, while the columns are separated by the semicolon operator (;). This fact can also be used in the case of multidimensional arrays:

```
1  -->a = 2:7
2  a =
3  2.   3.   4.   5.   6.   7.
4  -->A = [a;a]
5  A =
6  2.   3.   4.   5.   6.   7.
7  2.   3.   4.   5.   6.   7.
8  -->AA = [A,A]
9  AA =
10 2.   3.   4.   5.   6.   7.   2.   3.   4.   5.   6.   7.
11 2.   3.   4.   5.   6.   7.   2.   3.   4.   5.   6.   7.
```

3.3.12 Appending Rows and Columns

When an entire row or column of a matrix needs to be appended, only one thing must be taken into consideration. The size of the new matrix, which is used for this purpose, must match the row and column requirement. As an example, let's define an array A,B,D with sizes 2×2, 1×2, and 2×1, respectively. Row matrix B can be inserted as a row to A and column matrix D can be inserted as a row to A:

```
1  -->A = [4,-3;5,-5]
2  A =
3  4.    -3.
4  5.    -5.
5  -->B = [-9,1]
6  B =
7  -9.    1.
```

```
 8  --->size(A)
 9  ans =
10  2.    2.
11  --->size(B)
12  ans =
13  1.    2.
14  --->C = [A;B]
15  C =
16  4.    -3.
17  5.    -5.
18  -9.    1.
19  --->size(C)
20  ans =
21  3.    2.
22  --->D = [5;6]
23  D =
24  5.
25  6.
26  --->size(D)
27  ans =
28  2.    1.
29  --->E = [A,D]
30  E =
31  4.    -3.    5.
32  5.    -5.    6.
33  --->size(E)
34  ans =
35  2.    3.
```

3.3.13 Deleting a Row and/or Column of a Matrix

Rows and columns can be deleted by assigning null matrices [ ] to them. For example, ()1,:)=[ ] deletes the first row, and ():,1) deletes the first column of a matrix:

```
1   -->A = rand(3,3)
2   A =
3   0.88    0.93    0.36
4   0.65    0.21    0.29
5   0.31    0.31    0.57
6   -->A (1,:)=[]
7   A =
8   0.65    0.21    0.29
9   0.31    0.31    0.57
10  -->A(:,1)=[]
11  A =
12  0.21    0.29
13  0.31    0.57
```

3.3.14 Concatenation along a Dimension

Concatenation of two matrices along a dimension can be obtained using cat(dim, A,B,...) where dim presents the dimension and A and B are input matrices. The following example demonstrates its usage:

```
1   -->A = rand (2,2) // 2X2 matrix is defined
2   A =
3   0.48    0.59
4   0.33    0.5
```

```
5   --->B = rand(2,2)
6   B =
7   0.44   0.63
8   0.27   0.41
9   --->cat(1,A,B) // A and B are concatenated along columns
    (dimension=1)
10  ans =
11  0.48   0.59
12  0.33   0.5
13  0.44   0.63
14  0.27   0.41
15  --->cat(2,A,B) // A and B are concatenated along columns
    (dimesnion -2)
16  ans =
17  0.48   0.59   0.44   0.63
18  0.33   0.5    0.27   0.41
19  --->C = cat(3,A,B) // A and B are concatenated a long third
    dimesnion where first element is A and second is B
20  C =
21
22  (:,:,1)
23
24  0.48   0.59
25  0.33   0.5
26  (:,:,2)
27
28  0.44   0.63
29  0.27   0.41
30  --->size(C) // new matrix has 3 dimesnions i.e its size is
    2X2X2
31  ans =
32  2.   2.   2.
```

When cat(1,A,B) is fed to the command prompt, A and B are concatenated row-wise and cat(2,A,B) results concatenation columnwise. In the case of cat(3,A,B), a new matrix is created whose first element of the third dimension is the matrix A and the second element is the matrix B.

3.4 Logical Operations on Arrays

Logical operations are performed elementwise. They are mainly used for comparisons. Suppose we wish to find if all elements of two arrays are same. Here == is used. Similarly, <, ><=, and >= operators can also be used. It's important to note that these element-by-element operations produce an array of boolean values. In other words, the element result is either %T or %F, signifying a True or False value for the comparison:

```
1  -->a = [1,2,3;4,5,6]
2  a =
3  1.   2.   3.
4  4.   5.   6.
5  -->a == a // all elements are same in value
6  ans =
7  T T T
8  T T T
9  -->b = [1,2,3;4,5,8]
10 b =
11 1.   2.   3.
12 4.   5.   8.
13 -->a == b // all elements except row 2 column 3 is
   dis-similar
14 ans =
```

```
15  T T T
16  T T F
17  ——>b > a // all elements of b except row 2 column 3, are
    not bigger than those of a
18  ans =
19  F F F
20  F F T
21  ——>b > = a // all elements are either equal OR greater than
    corresponding elements
22  ans =
23  T T T
24  T T T
```

3.5 Automatic Generation of Vectors

Most often, users want to generate a long list of sequential numbers as elements of an array. This is called the *automatic generation of vectors.* You can generate a series of numbers and store them as arrays by using the following command:

start:step:stop

```
1  ——>a = 1:1:10 // start from 1 with a step of 1 and end at 10
2  a =
3  1.    2.    3.    4.    5.    6.    7.    8.    9.    10.
4  ——>a = [1:1:10]
5  a =
6  1.    2.    3.    4.    5.    6.    7.    8.    9.    10.
7  ——>a = 1:2:10 // star at 1 with step of 2 and end at 10
8  a =
9  1.    3.    5.    7.    9.
```

Please note that brackts[ ] are optional here. Also, if the step is not defined, its default value (1) is taken:

```
1  -->a = 1:10 // start at 1 and end at 10 (defaule step is 1)
2  a =
3  1.    2.    3.    4.    5.    6.    7.    8.    9.    10.
```

3.5.1 Linearly Spaced Vectors

The command linspace(start,stop,n) produces an array starting from the first number and stopping at the second one, with a total of n numbers. Hence, they are linearly spaced:

```
1  -->a = linspace(1,2,5)
2  a =
3  1 .     1.25    1.5    1.75    2.
4  -->a = linspace(1,10,10)
5  a =
6  1.    2.    3.    4.    5.    6.    7.    8.    9.    10.
```

3.5.2 Logarithmically Spaced Vectors

Similar to linspace, logspace(start,stop,n) produces n numbers from start to stop that are linearly spaced in logarithmic nature:

```
1  -->a = logspace(1,10,5)
2  a =
3  10.    1778.28    316228.    5.6D+07    1.0D+10
```

3.5.3 meshgrid

The built-in functions linspace and logpspace are used to derive points on a line segment graphically. If we wish to derive a surface as equally spaced mesh points, we use the built-in function meshgrid as follows:

```
1  -->a = 1:0.1:100;
2  -->b = 1:0.1:100;
3  -->[X, Y]= meshgrid(a,b);
```

Here [X,Y] is an 2D array containing two numbers for each element. One of the elements is taken from the array a and another element is taken from the array b.

3.5.4 ndgrid

The built-in function meshgrid provides a 2D grid of points on which a 3D function can be defined to fabricate a 3D surface. The built-in function ndgrid provides an n-dimensional grid in a similar fashion:

```
1  -->a = 1:0.1:10;
2  -->b = 1:0.1:10;
3  -->c = 1:0.1:10;
4  -->[X Y Z] = ndgrid(a,b,c);
```

3.6 Matrix Manipulations

Some common matrix manipulations have already been written in function form, which makes it easier for a developer to use them right away rather than investing time to write an optimum code.

3.6.1 Scaling a Matrix

Scalar multiplication of a matrix can be performed by simply multiplying a scalar with a matrix. It can be argued that a scalar is a 1×1 dimensional matrix. If so, how can it operate on, say, a $m \times n$ matrix? In this case, a scalar is projected onto a $m \times n$ dimension and then an elementwise operation is performed:

```
1  -->a = [1 2 3; 4 5 6; 7 8 9]
2  a =
3  1.    2.    3.
4  4.    5.    6.
5  7.    8.    9.
6  -->a 2 = 2*a
7  a2 =
8  2.    4.     6.
9  8 .   10.    12.
10 14.   16.    18.
11 -->a3 = a/2
12 a3 =
13 0.5   1.    1.5
14 2.    2.5   3.
15 3.5   4.    4.5
```

3.6.2 Reshaping a Matrix

The number of rows and columns can be changed provided that the total number of elements remains the same. For this purpose, the matrix() function can be used. Its usage is illustrated well in help("matrix").

```
1  -->a = [1 2 3; 4 5 6; 7 8 9] // 3X3 matrix of 9 elements
2  a =
3  1.    2.    3.
4  4.    5.    6.
5  7.    8.    9.
```

```
6   -->a1 = matrix(a,1,9) // 1X9 matrix of 9 elements
7   a1 =
8   1.    4.    7.    2.    5.    8.    3.    6.    9.
9   -->a1 = matrix(a,2,5) // cant make a 2X5 matrix with 9
    elements
10  Wrong size for argument: Incompatible dimensions.
```

3.7 Special Matrices

Matrix algebra defines some kinds of matrices that are special in nature and find their use in some problems. Octave has some functions defined to create these matrices.

3.7.1 Upper and Lower Triangular Matrices

An upper triangular matrix is one where only the diagonal and elements above the diagonal are non-zero. Similarly, a lower triangular matrix is one where the diagonal and the elements below the diagonal are non-zero. tril and triu fabricate a lower and upper triangular matrix:

```
1   -->a = rand(3,3)
2   a =
3   0.40409     0.31513     0.02929
4   0.41941     0.78735     0.18410
5   0.24369     0.63817     0.26223
6   -->tril(a)
7   ans =
8   0.40409     0.          0.
9   0.41941     0.78735     0.
10  0.24369     0.63817     0.26223
11  -->triu(a)
```

```
12  ans =
13  0.40409   0.31513     0.02929
14  0.          0.78735     0.18410
15  0.          0.          0.26223
```

3.7.2 Ones and Zeros Matrices

A matrix having all its numbers as 1 or 0 makes up a ones and zeros matrix, respectively:

```
1  -->ones(3,3)
2  ans =
3  1.    1.    1.
4  1.    1.    1.
5  1.    1.    1.
6  -->zeros(3,3)
7  ans =
8  0.    0.    0.
9  0.    0.    0.
10 0.    0.    0.
```

3.7.3 Diagonal Matrices

Diagonal matrices with predefined diagonal elements can be fabricated using the diag() command:

```
1  -->diag([1 3])
2  ans =
3  1.    0.
4  0.    3.
5  -->diag([1 3 4 6 7])
```

```
 6  ans =
 7  1.    0.    0.    0. ·   0.
 8  0.    3.    0.    0.    0.
 9  0.    0.    4.    0.    0.
10  0.    0.    0.    6.    0.
11  0.    0.    0.    0.    7.
```

3.7.4 Special Matrices

The built-in function testmatrix() generates special matrices like a magic square, Franck, and the inverse of a $n \times n$ Hilbert matrix when used with input arguments magi, frk, and hilb as strings:

```
 1  -->testmatrix('magi',3)
 2  ans =
 3
 4  8.    1.    6.
 5  3.    5.    7.
 6  4.    9.    2.
 7
 8  -->testmatrix('frk',4)
 9  ans =
10
11  4.    3.    2.    1.
12  3.    3.    2.    1.
13  0.    2.    2.    1.
14  0.    0.    1.    1.
15
16  -->testmatrix('hilb',4)
```

```
17  ans =
18
19     16.    −120.     240.    −140.
20   −120.    1200.   −2700.    1680.
21    240.   −2700.    6480.   −4200.
22   −140.    1680.   −4200.    2800.
```

3.8 Mathematical Matrix Operations

A variety of matrix operations like dot product, cross product, and so on, exists in matrix algebra. These operations can be performed in Scilab using appropriate code for the same.

3.8.1 Dot Products

A dot product of two vectors produces a scalar as follows:

$$\vec{A_1} = x_1\vec{i} + y_1\vec{j} + z_1\vec{k}$$

$$\vec{A_2} = x_2\vec{i} + y_2\vec{j} + z_2\vec{k}$$

$$\vec{A_1} \cdot \vec{A_2} = x_1 \times x_2 + y_1 \times y_2 + z_1 \times z_2$$

This operation can be performed using the following code where the vector dot product is defined as multiplication of a with transpose of b:

```
1  --->a1=[1 2 3];
2  --->a2=[2 3 4];
3  --->a1*a2'
4  ans =
5  20.
```

3.8.2 Cross Products

The built-in function cross() returns the cross product of two input vectors:

```
1  -->a=[1 2 3];
2  -->b=[2 3 4];
3  -->cross(a,b)
4  ans =
5  - 1.    2.    -1.
6  -->a1=[%t,%t,%f];
7  // One can define vector using booleans too
8  -->a2=[%f,%f,%f];
9  -->cross(a1,a2)
10 ans =
11 0.    0.    0.
12 -->cross(a1,a) // cross product of boolean vector with
   vector of real numbers
13 ans =
14 3.    - 3.    1.
```

3.9 Discrete Mathematics

Scilab has a limited but useful set of built-in functions to work with discrete mathematics as follows:

1. primes outputs the primes numbers until the given number

    ```
    1  -->x = 30;
    2  -->y = primes(x)
    3  y =
    4  2.  3.  5.  7.  11.  13.  17.  19.  23.  29.
    5
    ```

2. factor derives the factors of a given number—if *n* is the given number, then factor produces an array of prime numbers a,b,c ...z such that

$$n = \prod(a,b,c...z)$$

```
1    -->x=3*10e4;
2    -->y=factor(x)
3    y =
4    2.   2.   2.   2.   2.   3.   5.   5.   5.   5.   5.
5    -->x=79867858;
6    -->y=factor(x)
7    y =
8    2.      7.      5704847.
9    -->x=9999;
10
11   -->y=factor(x)
12   y =
13   3.      3.      11.      101.
14
```

3. rat() derives a floating point rational representation of a given number. help rat gives its detailed usage for input arguments.

```
1    -->[n,d]=rat(%pi)
2    d =
3    113.
4    n =
5    355.
6    -->[n,d]=rat(%e)
7    d =
```

```
 8   465.
 9   n =
10   1264.
11   -->[n,d]=rat(500)
12   d =
13   1.
14   n =
15   500.
16   -->[n,d]=rat(500.5)
17   d =
18   2.
19   n =
20   1001.
21
```

4. factorial outputs the factorial of a given number

```
 1   -->x = 87;
 2   -->y=factorial(x)
 3   y =
 4   2.10D+132
 5   -->x = 7;
 6   -->y=factorial(x)
 7   y =
 8   5040.
 9   -->x=0;
10   -->y=factorial(x)
11   y =
12   1.
13
```

5. perm outputs **permutations** of a given set of numbers

```
1    -->x = [1 2 3]
2    x =
3    1.    2.    3.
4    -->y=perms(x)
5    y =
6    3.    2.    1.
7    3.    1.    2.
8    2.    3.    1.
9    2.    1.    3.
10   1.    3.    2.
11   1.    2.    3.
12
```

3.10 Finding Roots for Sets of Linear Equations

As an example to practically apply arrays to solve real-world problems, let's use arrays to find roots of a set of linear equations. Let's assume that we have the following:

$$2x + 4y - 3z = 4 \tag{3.1}$$

$$-2x - 3y + 2z = -3 \tag{3.2}$$

$$4x + 6y - 8z = 1 \tag{3.3}$$

We wish to find those values of x, y, and z for which all three equations hold true. To do the same, the first matrix form of these equations can be written as follows:

$$\begin{bmatrix} 2 & 4 & -4 \\ -2 & -3 & 2 \\ 4 & 6 & -8 \end{bmatrix} \times \begin{bmatrix} x \\ y \\ z \end{bmatrix} = \begin{bmatrix} 4 \\ -3 \\ 1 \end{bmatrix} \tag{3.4}$$

If we assume that

$$A = \begin{bmatrix} 2 & 4 & -4 \\ -2 & -3 & 2 \\ 4 & 6 & -8 \end{bmatrix} \tag{3.5}$$

$$X = \begin{bmatrix} x \\ y \\ z \end{bmatrix} \tag{3.6}$$

$$B = \begin{bmatrix} 4 \\ -3 \\ 1 \end{bmatrix} \tag{3.7}$$

then we can write

$$Ax = B \tag{3.8}$$

for which the solution is

$$X = A^{-1}B \tag{3.9}$$

This can be found with ease in Scilab using just a one-line command:

```
1  --->A = [2,4,-4;-2,-3,2;4,6,-8]
2  A =
3
4  2.    4.    -4.
5  -2.   -3.    2.
6  4.    6.    -8.
7
8
9  --->B = [4; - 3;1]
10  B =
11
12  4.
13  -3.
14  1.
15
16  --->A\B
17  ans =
18
19  -2.5
20  3.5
21  1.25
```

Thus, $x = -2.5$, $y = 3.5$, and $z = 1.25$ satisfy the equations. In this way, Scilab can help perform complex matrix calculations with ease.

3.11 Summary

Array-based computing lies at the very heart of modern computational techniques. Scilab presents a very suitable platform to perform these techniques with ease. A variety of predefined functions enable users to save time while prototyping a problem. Flexible methods to define multidimensional arrays and performing fast computation are the main necessitites of our times. Most of the time spent during a simulation is either in loops or in array operations. Predefined array operations have been optimized with algorithms for reliability, time efficiency, and effective memory management.

CHAPTER 4

Plotting

4.1 Introduction

Without visualization, numerical computations are difficult to understand and eventually judge. Producing publication-quality images of complex plots that give a meaningful analysis of numerical results has been a challenge for scientists all over the world. Many commercial softwares satisfy this need. Scilab also provides this facility quite efficiently. Its plotting features includs choosing from various types of plots in 2D and 3D regimes; decorating plots with additional information such as titles, labeled axes, grids, and data labels; and writing equations and other important information about the data. The following sections will describe these actions in detail. It is worth mentioning that plotting capabilities are essential to certain numerical analysis experiments since visual directions from the progressive steps give an intuitive understanding of the problem under consideration.

4.2 2D Plotting

4.2.1 plot(x,y)

Since we need data on two axes to be plotted, we first need to create them. Let's assume that the x-axis has 100 linearly spaced data points and the points on the y-axis are defined by an equation, as illustrated in Figure 4-1:

$$y = x^2$$

© Sandeep Nagar 2017
S. Nagar, *Introduction to Scilab*, https://doi.org/10.1007/978-1-4842-3192-0_4

```
1  >> x = linspace(0,100,100);
2  >> y = x.^2
3  >> plot(x,y)
```

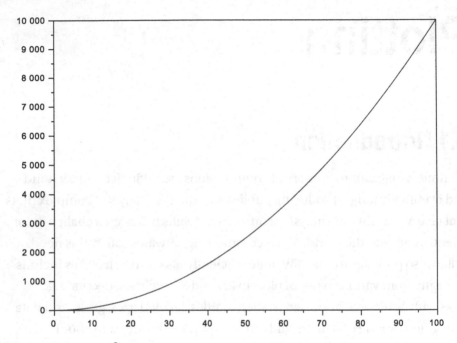

Figure 4-1. $y = x^2$

First, we define a variable x and place 100 equally spaced data points from 0 to 100. This makes a 1 x 100 matrix. Using a scalar operation of exponentiation, we define a variable y as x^2. It is important to note that this operation is defined using an elementwise operator so that each element of matrix x is operated by the operator. Without this approach, the array would have to be squared, that is, multipled by itself. This would produce an error since an $n \times m$ matrix can only be multipled by a $m \times n$ matrix. Finally, one can use the function plot(), which takes two arguments as the x-axis and y-axis data points.

Typing help plot or help'plot' at the command prompt gives useful insight into this wonderful function written to plot two-dimensional data. By default, successive plots are superposed. To clear the previous plot, use clf(). Also, the plot function can evaluate the input arguments:

```
1  -->plot(x = linspace(0,100,100),x.^2.5,'r*')
```

The result of this line code is shown in Figure 4-2. Please note that the third argument, a string: "r*", plots the data with * at each point of input array.

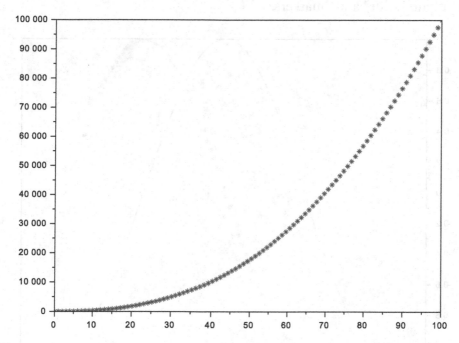

Figure 4-2. $y = x^2$

Multiple plots can be plotted using the plot() command. As an example, let's plot *sin(x)* and *cos(x)* functions between $-\pi$ and π. The following is the first version of the code:

```
1  -->x = linspace(-%pi,%pi,100);
2  -->plot(x,sin(x),'r*',x,cos(x),'b-')
```

The following code is the second version of the code:

```
1  -->x = linspace(-%pi,%pi,100);
2  -->plot(x,sin(x),'r*')
3  -->plot(x,cos(x),'b-')
```

Both codes produce similar figures because, in the second version, the second plot is overwritten on the first plot to produce multiple plots on the same figure window. Figure 4-3 shows the result where *sin(x)* and *cos(x)* are plotted with * and - markers.

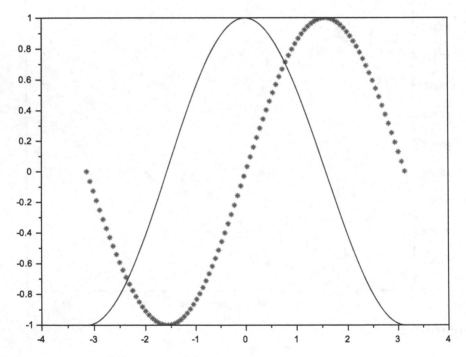

Figure 4-3. *sin(x) and cos(x)*

4.2.2 plot2d(), plot2d2(), plot2d3(), and plot2d4()

The plot2d command works similarly to the plot command. In addition, some other variations of these functions exist with respect to specialized way in which plots are plotted:

1. plot2d2 same as plot2d, but the curve is supposed to be piecewise constant

2. plot2d3 same as plot2d, but the curve is plotted with vertical bars

3. plot2d4 same as plot2d, but the curve is plotted with arrows

Listings 4-1, 4-2, 4-3, and 4-4 illustrate their usage. Figures 4-4 and 4-5 show how they are represented graphically.

Listing 4-1. plot2d.sce

```
1  // Program to plot using plot2d function
2  x = linspace(-%pi,%pi,20)
3  plot2d(x,sin(x))
4  xtitle('Graph for plotting sin(x) using plot2d')
5  xlabel('angle')
6  ylabel('sin(x)')
```

Listing 4-2. plot2d2.sce

```
1  // Program to plot using plot2d2 function
2  x = linspace(-%pi,%pi,20)
3  plot2d2(x,sin(x))
4  xtitle('Graph for plotting sin(x) using plot2d2')
5  xlabel('angle')
6  ylabel('sin(x)')
```

Listing 4-3. plot2d3.sce

```
1  // Program to plot using plot2d3 function
2  x = linspace(-%pi,%pi,20)
3  plot2d3(x,sin(x))
4  xtitle('Graph for plotting sin(x) using plot2d3')
5  xlabel('angle')
6  ylabel('sin(x)')
```

Listing 4-4. plot2d4.sce

```
1  // Program to plot using plot2d4 function
2  x = linspace(-%pi,%pi,20)
3  plot2d4(x,sin(x))
4  xtitle('Graph for plotting sin(x) using plot2d4')
5  xlabel('angle')
6  ylabel('sin(x)')
```

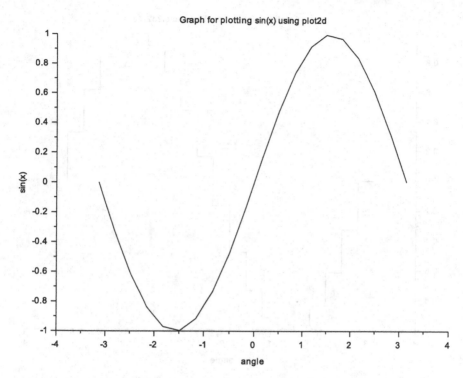

Figure 4-4. *sin(x) and cos(x)*

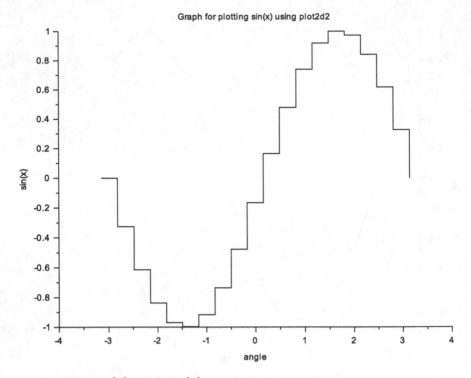

Figure 4-5. *sin(x) and cos(x)*

4.2.3 polarplot()

During mathematical analysis, polar coordinates become important in cases that do not exhibit symmetry in Cartesian systems, but that show symmetry in polar coordinates. In these cases, we prefer to plot in polar coordinates rather than Cartesian coordinates. So, instead of *x, y* our coordinates are *r, θ*, which are related by equations, as shown in Figures 4-6 and 4-7.

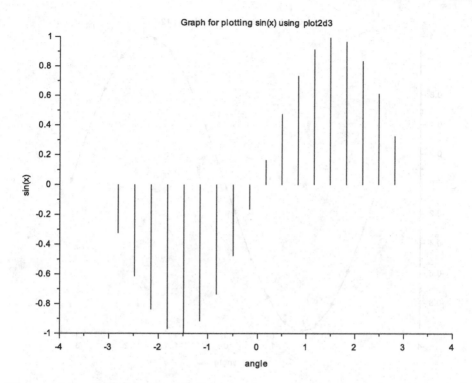

Figure 4-6. *sin(x) and cos(x)*

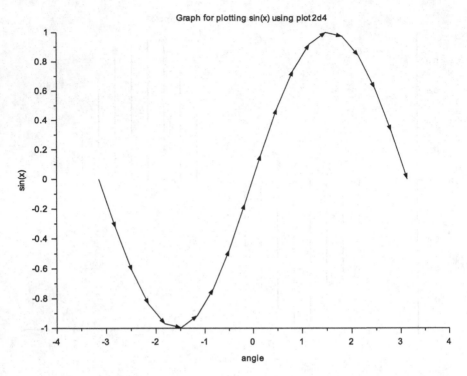

Figure 4-7. *sin(x) and cos(x)*

$$x = r \times cos(\theta)$$

$$y = r \times sin(\theta)$$

The code given by polarplot.sce plots one such case, as shown in Listing 4-5. The result is given in Figure 4-8.

Listing 4-5. polarplot.sce

```
1  // Program to plot using polarplot function
2  theta= 0:.01:3*%pi;
3  polarplot(sin(2*theta), cos((2*theta)))
4  xtitle('Using polarplot')
```

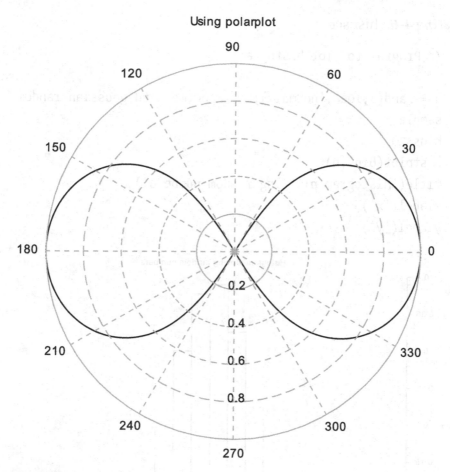

Figure 4-8. *Polar graph*

4.3 Special Plots

4.3.1 Histograms

Histograms are traditionally used to visualize the number of events occurring within different ranges. Scilab provides a useful function, histplot(), for this purpose that takes input for the number of bins and data provided as an array. For example, in the code titled hist.sce found in Listing 4-6, the normalized distribution of random numbers is plotted in Figure 4-9.

Listing 4-6. hist.sce

```
1  // Program to plot histogram
2
3  n = rand(1,10e4,'normal');   // normalized gaussian random
   sample
4  bins=20
5  histplot(bins,n)
6  title('Histogram plotting random numbers')
7  xlabel('x')
8  ylabel('y')
```

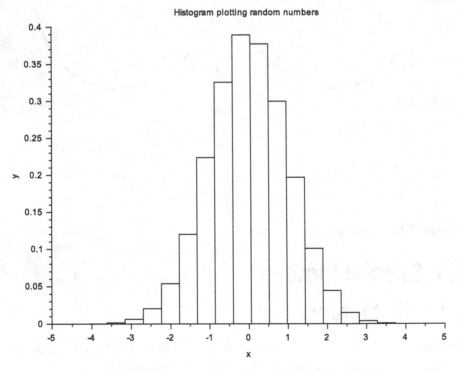

Figure 4-9. *Histogram plot*

Histograms are used to show the sampled data. From a dataset, one can segregate the data points according to their values into the number of bins and visualize which range of values dominate the profile. This provides useful information about the system. For example, values following normalized distribution represent systems that always follow certain kinds of statistics. In the present case depicted in Figure 4-9, most of the data events occur in the middle. If this graph represents the probability of finding an electron between two atoms, then we can conclude that the electron is found mostly in the middle of the atoms. If this graph represents grades obtained by students in a class, then we can deduce that very few students obtain either very good or very bad grades (extremes of graph) and that most of the students got close to grades of 50 percent.

4.3.2 matplot

To plot colors representing certain values, the matplot command is used, as shown in Listing 4-7 and illustrated in Figure 4-10.

Listing 4-7. matplot.sce

```
1  // Program to plot using Matplot function
2  x = [1 2 3 4;5 4 3 6]
3  Matplot(x)
4  grayplot(x,y,m)
5  xtitle('Using Matplot')
6  xlabel('x')
7  ylabel('y')
```

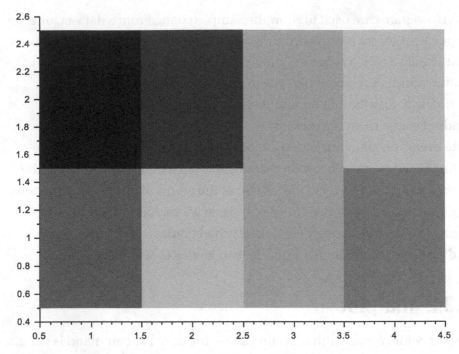

Figure 4-10. *Matplot figure for [1 2 3 4;5 4 3 6]*

4.3.3 grayplot

Sometimes, we need to see the value of a two-dimensional function on a 2D plot. This is done by visualizing the components of the functional value *z* on *x* and *y* coordinates. grayplot plots a graph window for a 2D plot of the surface given by *z* on a grid defined by *x* and *y*. Each rectangle on the grid is filled with a gray or color level depending on the average value of z on the corners of the rectangle. (See Listing 4-8 and Figure 4-11.)

Listing 4-8. grayplot.sce

```
1  // Program to plot using grayplot function
2  x = −10:10;
3  y = −10:10;
4  m = rand(21,21);
```

```
5  grayplot(x,y,m)
6  xtitle('Using grayplot for random numbers')
7  xlabel('x')
8  ylabel('y')
```

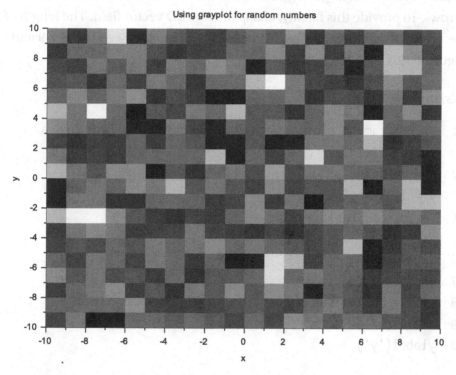

Figure 4-11. *grayplot figure for random numbers*

Figure 4-11 shows the color value as per the value of z (which is given by a random number in this case). The graph will result in different plots each time it is run because the value z is given by a random number. In fact, the effectiveness of random-number-generating algorithms can be visually inspected using grayplot by checking if two colors are situated close to each other.

4.3.4 champ

Scientific and engineering studies employ the theoretical study of vector fields. Both in theoretical and practical studies, it is sometimes important to plot vector fields as per a given equation. Vectors are best described by arrows. To provide this facility, champ draws a 2D vector field. The length of the arrows is proportional to the intensity of the field. (See Listing 4-9 and Figure 4-12.)

Listing 4-9. champ.sce

```
1   // Program to plot using champ function for plotting a
    vector field
2   x = linspace(-1,1,10);
3   y = linspace(-1,1,10);
4   [X,Y] = meshgrid(x,y);
5   fy = 3.*Y;
6   fx = 0.5.*X;
7   champ(x,y,fx',fy')
8   xtitle('Using champ function to plot vector field')
9   xlabel('x')
10  ylabel('y')
```

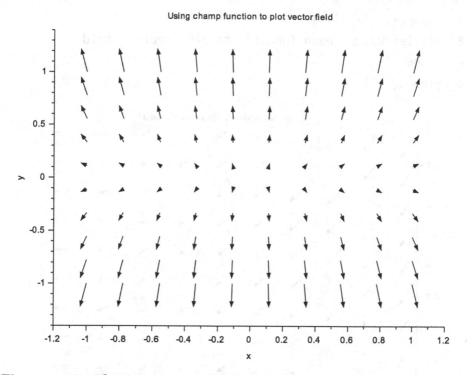

Figure 4-12. *Plotting a vector field using champ*

Complex polynomials can be used to represent the horizontal and vertical components of the vector field, which are usually measured. champ provides an easy way to visualize the vector field and then make deductions about forces arising out of it. (See Listing 4-1 and Figure 4-13.)

Listing 4-10. champ1.sce

```
1  // Program to plot using champ function for plotting a
   vector field
2  x = linspace(-2,2,10);
3  y = linspace(-2,2,10);
4  [X,Y] = meshgrid(x,y);
5  fy = X.^3-2.*X.^2+4.*X-10;
6  fx = X.^3+2.*X.^2-4.*X+10;
```

```
7  champ(x,y,fx',fy')
8  xtitle('Using champ function to plot vector field')
9  xlabel('x')
10 ylabel('y')
```

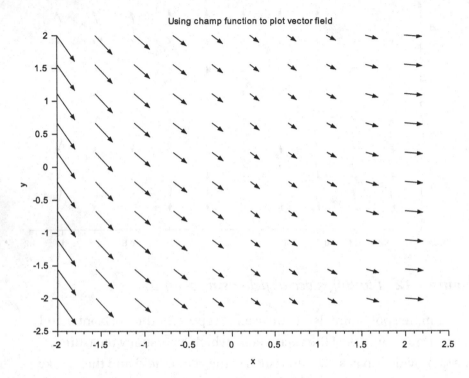

Figure 4-13. *Plotting a vector field using champ*

Adding color (indicating field intensity) to the arrows gives a more meaningful visual presentation, as shown in Listing 4-11 and Figure 4-14.

Listing 4-11. champcolor.sce

```
1  // Program to plot using champ function for plotting a
   vector field
2  x = linspace(-1,1,20);
3  y = linspace(-1,1,20);
```

```
4   [X,Y] = meshgrid(x,y);
5   fy = 3.*Y;
6   fx = 0.5.*X;
7   champ1(x,y,fx',fy')
8   xtitle('Using champ function to plot vector field')
9   xlabel('x')
10  ylabel('y')
```

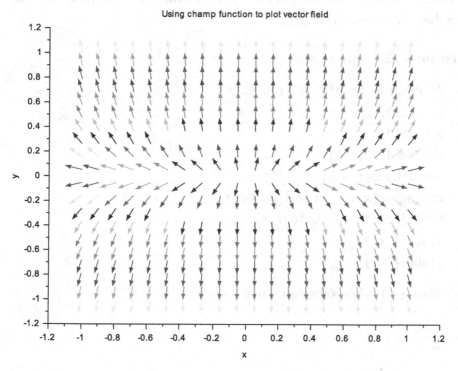

Figure 4-14. *Plotting a vector field using champ*

4.3.5 Contour Maps

Contour maps are frequently used to visualize the projection of a function on a 2D surface. They are used in various branches of science and engineering, especially geophysics where information of measurement parameters like temperature, pressure, and humidity is projected onto a map. In addition, contour maps are also used to visualize the 2D mathematical projection on a plane.

Scilab provides the built-in function `contour2d()` for this purpose. Its usage is explained in the example code in Listing 4-12. Detailed usage of the function can be learned by using the command `help contour2d`. Essentially, the function takes x, y, and z values and the value for the number of levels. The code produces random numbers and plots them as per their value on a 2D matrix.

Listing 4-12. contour.sce

```
1  // Program to plot 2D contours
2  x = 1:10;
3  y = 1:10;
4  z = rand(10,10);
5  level_number = 8;
6  contour2d(x,y,z, level_number)
7  title('Plotting contour map')
8  xlabel('x')
9  ylabel('y')
```

Figure 4-15 shows the result of the code.

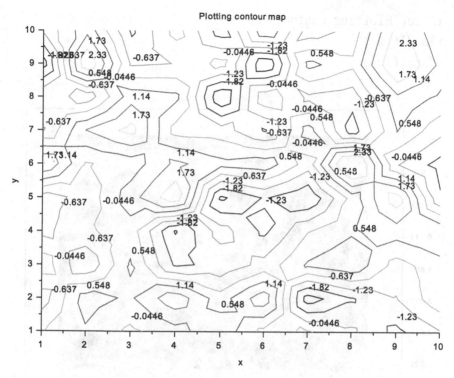

Figure 4-15. *Plotting a vector field using champ*

Instead of contour outlines, **filled contours** can be obtained using the built-in function contourf(). It follows a similar syntax as the contour2d() function. (See Listing 4-13 and Figure 4-16.)

Listing 4-13. contourf.sce

```
1  // Program to plot 2D contours
2  x = 1:10;
3  y = 1:10;
4  z = rand(10,10);
5  level_number = 8;
6  contourf(x,y,z, level_number)
```

113

```
7  title('Plotting contour map')
8  xlabel('x')
9  ylabel('y')
```

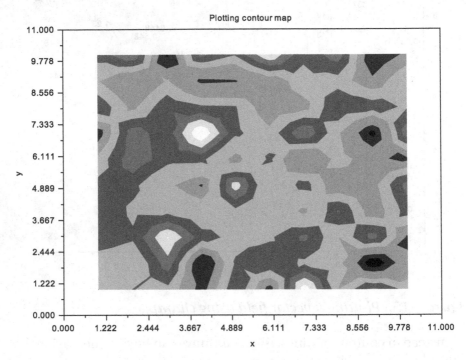

Figure 4-16. *Plotting a vector field using champ*

The color-coded areas are on the same level in Figure 4-16. Since the z values are taken from a random number generator function, `rand()`, the figure will be different each time the code `contourf.sce` is run.

Contour maps are easy way to visualize the smoothness of 2D data. Smooth data will have little variation and, hence, will fall under the same level. The choice of the level range will define the degree of smoothness. For example, suppose an infrared camera measures the temperature of a room and the data are color-coded for visual display. Contours will be defined by the different levels of temperature ranges. If the temperature is uniform in a region, it will be coded with a similar color. The difference in temperature of that region will be within the range defined and, thus, the definition of the range will become critical. If the minimum and maximum values of temperature are T_1 and T_2 and the number of levels are n, then the temperature range of a zone is given by

$$\frac{|T_2 - T_1|}{n}$$

4.4 2D Animation

During simulation, it is sometimes useful to animate an equation to understand its evolution. Scilab contains very useful functions for this purpose. Please note that figures provided are snapshots-at-the-end for the animations. Users should write the code as discussed and test it on their own to see the animation on a computer monitor. Animations are found to be particularly useful for teaching concepts to students.

4.4.1 comet

The comet() function comes in handy in such cases. In Listing 4-14, we animate the equation $x^5 - x^3 + x + 5$ using the comet function.

Listing 4-14. comet.sce

```
1  // Program to demonstarte usage of comet
2
3  x = linspace(-%pi,%pi,500);
4  comet(x,%s^5-%s^3+%s+5)
5  xtitle('Using comet function to animate an equation')
6  xlabel('x')
7  ylabel('y')
```

4.4.2 paramfplot2d()

Time evolution of a function at multiple points can be represented by the
paramfplot2d() function. Let's consider the evolution of the function
$y = sin(x) + sin(2x) + sin(4x)$ (mixing of a signal with double and four times
its own frequency). The resulting signal is the complex waves combining
these three primary waves *(sin(x), sin(2x),* and *sin(4x))*, as shown in
Figure 4-17 and Listing 4-15.

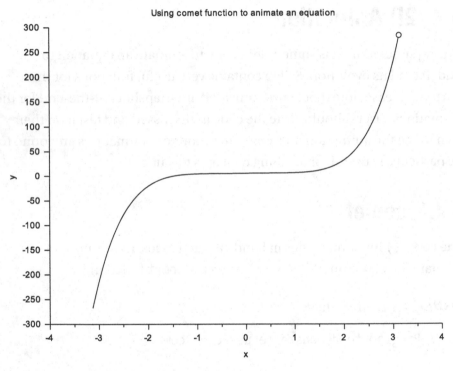

Figure 4-17. *Plotting using the comet function*

Listing 4-15. paramfplot2d.sce

```
1  deff('y=f(x,t)','y=t*sin(x)+sin(2*x)+sin(4*x)')
   // Defining equation
2  x = linspace(-4*%pi,4*%pi,10e3); // defining spatial steps
3  t = 100;
4  time_speed = 0:0.05:1; // step size determines speed of
   animation
5  paramfplot2d(f,x,time_speed);
6  xtitle('Using paramfplot2d function to animate an equation')
7  xlabel('x')
8  ylabel('y')
```

Please note that deff is used to define a function. Its usage can be checked from help deff. Its detailed usage will be discussed in the chapter concerning the definition of Scilab function.

4.5 Plotting Multiple Plots in the Same Graph

Multiple plots can be plotted within the same figure by simply supplying x-axis and y-axis vectors, as shown in Listing 4-16 and Figure 4-18.

Listing 4-16. multi.sce

```
 1  // Program for multiple plots with legends
 2
 3  x = linspace(1,10,30);
 4  plot(x,x.^2,'k*')
 5  plot(x,x.^2.5,'r-')
 6  plot(x,x.^3,'b--')
 7  legend(['x^2';'x^3';'x^4']);
 8  title('Plotting multiple plots in same window')
 9  xlabel('x')
10  ylabel('y')
```

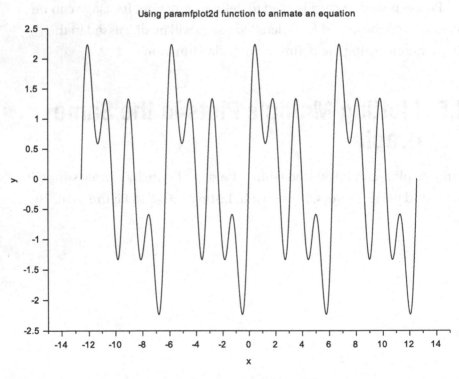

Figure 4-18. *Plotting using the paramfplot2d function*

Figure 4-18 is obtained by running the code in Listing 4-16. These types of plots are generally used to check the variation of results by varying a particular parameter.

4.5.1 Plotting Multiple Plots Separately

The subplot(row,coloumn, index) command is used to plot multiple plots within the same figure separately. subplot(2,2,4) means that the plot will be on the second row, second column, and fourth index. (See Listing 4-17 and Figure 4-19.)

Listing 4-17. subplot.sce

```
1  // Program to show usage of subplot function
2  // subplot function produces a figure as a matrix
3
4  x = linspace(−2*%pi,2*%pi,1000);
5  //1st figure of a 2X2 figure matrix
6  subplot(221)
7  plot(x,sin(x),'r*')
8  title('Plot for sin(x)')
9  xlabel('x')
10 ylabel('sin(x)')
11
12 //2nd figure of a 2X2 figure matrix
13 subplot(222)
14 plot(x,sin(x)+sin(2.*x),'b*')
15 title('Plot for sin(x) + sin(2x)')
16 xlabel('x')
17 ylabel('sin(x) + sin(2x)')
18
```

```
19  //3rd figure of a 2X2 figure matrix
20  subplot(2,2,3)
21  plot(x,sin(x) + sin(3.* x),'g*')
22  title('Plot for sin(x) + sin(3x)')
23  xlabel('x')
24  ylabel('sin(x) + sin(3x)')
25
26  //4th figure of a 2X2 figure matrix
27  subplot(2,2,4)
28  plot(x,sin(x) + sin(4.*x),'k*')
29  title('Plot for sin(x) + sin(4x)')
30  xlabel('x')
31  ylabel('sin(x) + sin(4x)')
```

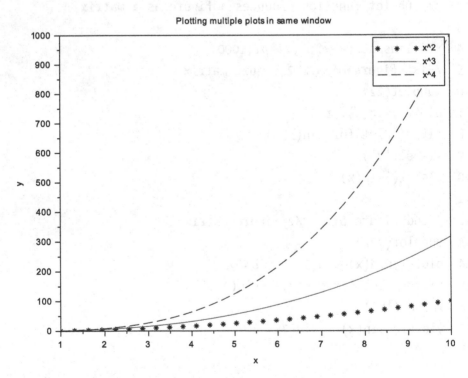

Figure 4-19. *Multiple plots within the same figure*

As seen in Figure 4-20, plots are organized as a matrix where the row number as well as the column number dictate its position. The index of the plot can then be used to treat it as an object for further processing on a graphical object. Theoretically, the limit of the number of rows and columns depends on the memory of computer but, while printing, users must understand that according to the size of plots, some of them will go outside the print area. Hence, judicious usage of this command is recommended. The automatic generation of a figure matrix can be achieved by generating the index of plots using a loop.

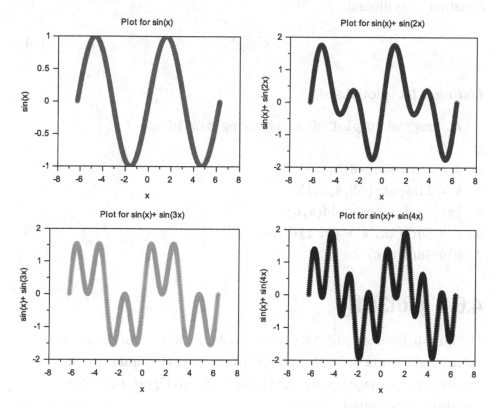

Figure 4-20. *Separate multiple plots within the same figure using a subplot*

4.6 3D Plots

There are various functions available for 3D plotting in Scilab. Choosing one of them depends on your particular problem.

4.6.1 plot3d

The built-in function plot3d() plots a 3D plot for the given values of x, y, and z defined as arrays. One of its uses is shown in Listing 4-18, where the equation 4.1 is plotted

$$z = \sqrt{x^2 + y^2} \tag{4.1}$$

Listing 4-18. plot3d.sce

```
1  // Program to plot 3D graph using plot3d
2
3  a = linspace(−8,8,41)';
4  b = linspace(−8,8,41)'
5  [xx, yy] = meshgrid(a,b);
6  c = sqrt(xx.^2 + yy.^2)+%eps;
7  plot3d(a,b,c)
```

4.6.2 plot3d1()

The built-in function plot3d1() plots a colored 3D plot for the given values of x, y, and z defined as arrays. The color of cells indicates the z value range. Its usage is shown in Listing 4-19 and Figure 4-21, where Equation 4.1 is plotted.

Listing 4-19. plot3d1.sce

```
1  // Program to plot 3D graph using plot3d1
2
3  a = linspace(-8,8,41)';
4  b = linspace(-8,8,41)'
5  [xx, yy] = meshgrid(a,b);
6  c = sqrt(xx.^2 + yy.^2)+%eps;
7  plot3d1(a,b,c)
```

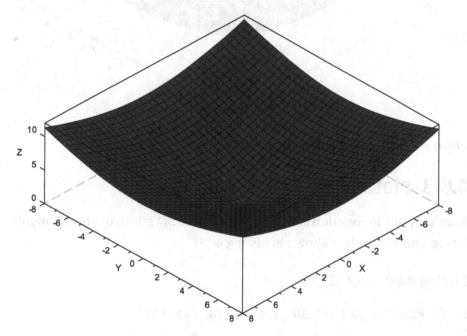

Figure 4-21. *3D plot of an equation*

The color coding provided by plot3d1() becomes an additional feature in some critical analyses. The contours in a 3D plot give the shape of projection areas where the value of the measured system is the same. For example, suppose Figure 4-22 represents the electric field near a charged particle placed at its center. The color represents the zone in 3D where the force on that charged particle will be within a certain range.

123

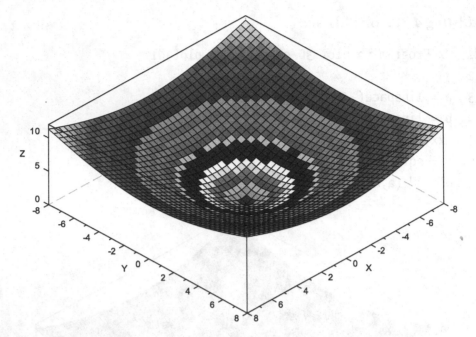

Figure 4-22. *3D plot of an equation*

4.6.3 plot3d2()

Another built-in function `plot3d2()` builds the 3D graph using rectangular facets. One example is shown in Listing 4-20.

Listing 4-20. plot3d2.sce

```
1  // Program to plot 3D graph using plot3d2
2
3  u = linspace(-%pi/2,%pi/2,40);
4  v = linspace(0,2*%pi,20);
5  X = cos(u)'*cos(v);
6  Y = cos(u)'*sin(v);
7  Z = sin(u)'*ones(v);
8  plot3d2(X,Y,Z);
```

The result is shown in Figure 4-23. It is important to note that a finer mesh will make the figure have better resolution of its 3D feature. For a finer mesh, the linspace() arguments should be defined with a smaller step size. However, at the same time, having a finer mesh increases the computational time. Thus, 3D rendering is a computationally intensive task, but it is sometimes desired. The choice of parameters depends upon the requirements of data analysis. When a finer resolution of 3D surface features is sought, users have to go for higher rendering power (which directly results in high computational power requirements).

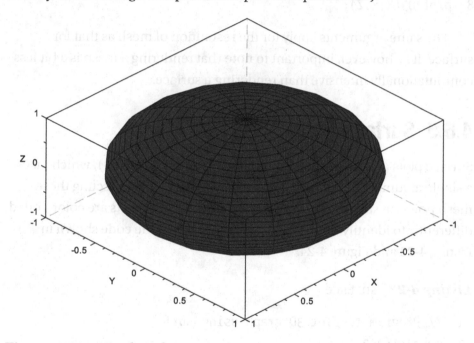

Figure 4-23. *3D plot of an equation*

4.6.4 plot3d3()

Another built-in function, plot3d3(), builds the 3D mesh graph using rectangular facets. The only difference between plot3d2 and plot3d3 is that the former produces a surface plot and the latter produces a mesh plot on rectangular facets. One example is shown in Listing 4-21.

Listing 4-21. plot3d3.sce

```
1  // Program to plot 3D graph using plot3d3
2
3  u = linspace(-%pi/2,%pi/2,40);
4  v = linspace(0,2*%pi,20);
5  X = cos(u)'*cos(v);
6  Y = cos(u)'*sin(v);
7  Z = sin(u)'*ones(v);
8  plot3d3(X,Y,Z);
```

The same arguments apply for the resolution of mesh as that for surface. It is, however, important to note that rendering a mesh is a bit less computationally intensive than rendering a surface.

4.6.5 Surface Plots

Surface plots are obtained using the built-in function surf(), which takes a single argument of the height of the mesh point. By connecting these mesh points, a surface is created. By default, surface facets are color-coded differently to identify them. This can be checked in the code shown in Listing 4-22 and Figure 4-24.

Listing 4-22. surf.sce

```
1  // Program to plot 3D graph using surf
2  subplot(2,2,1)
3  z = rand(10,10);
4  surf(z)
5  title('surf')
6
7  subplot(2,2,2)
8  surf(z,'facecol','red','edgecol','blu')
```

```
 9  title('surf function with face and edge color')
10
11  subplot(2,2,3)
12  surf(z,'facecol','interp')
13  title('surf function interpolated')
14
15  subplot(2,2,4)
16  x=rand(10,10);
17  y=rand(10,10);
18  z=rand(10,10);
19  surf(z,'facecol','red','edgecol','blu')
20  title('surf function with each coordinated defined
    seperately')
```

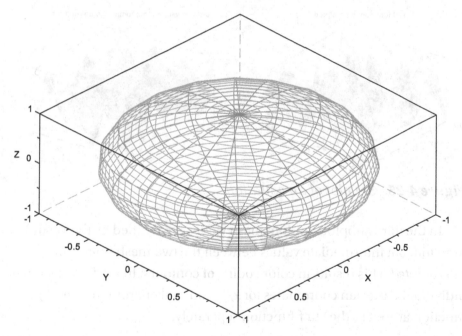

Figure 4-24. *3D plot of an equation*

Note that when additional arguments about *face color* being *red* and *edge colors* being *blue* are fed, the plot is plotted with a red-colored surface having blue-colored edges, as shown in Figure 4-25. The input to the surf() command is an array. It is created by the built-in function rand() in this case, but it can be created externally and fed to the surf() function.

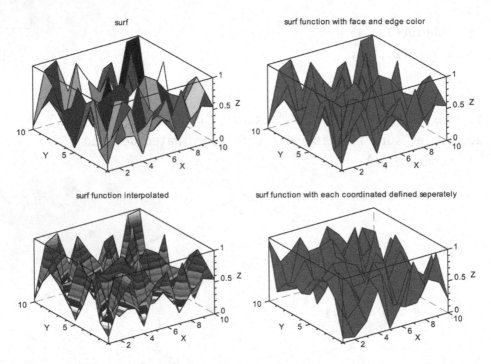

Figure 4-25. *3D plot of an equation*

In the third subplot, random mesh values are defined by the rand() function, but intermediate values between the two mesh points are *interpolated*. This results in color coding of contours. In the fourth subplot, individual Cartesian coordinates for *x, y,* and *z* taken from the rand() function are fed to the surf function seperately.

4.6.6 Mesh Plots

Mesh plots are generated by joining mesh points defined by coordinates. As an example, consider the code in Listing 4-23.

Listing 4-23. mesh.sce

```
1   // Program to explain the usage of mesh function
2
3   [X,Y] = meshgrid(-5:0.5:5, -5:.5:5);
4   Z = sin(X)+cos(Y);
5   mesh(X,Y,Z);
```

Here, a 2D coordinate system is generated and stored in two arrays named X and Y. For these points defining x and y, an equation is defined as follows:

$$z = sin(x) + cos(x) \tag{4.2}$$

These three values (one each from arrays x, y, and z) are fed to the mesh function.

The result is shown in Figure 4-26. The three coordinates (x, y, and z) are defined by the corresponding elements of vectors X, Y, and Z. These points are used to create the mesh plot.

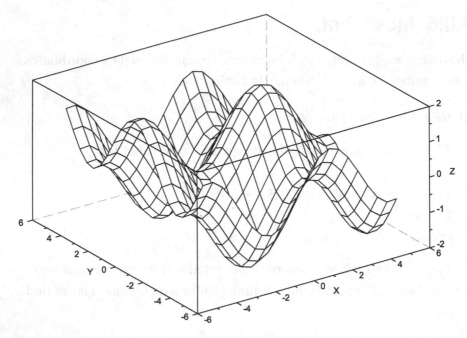

Figure 4-26. *3D plot of an equation*

4.6.7 3D Histogram

Just like 2D histograms, 3D histograms give visual clues about segregation of data having 2D dependency. 3D visualization is presented as vertical solid bars, as demonstrated in Listing 4-24.

Listing 4-24. hist3d.sce

```
1  // Program to make a 3D histogram
2
3  hist3d(5*rand(10,10));
4  hist3d(-5*rand(10,10));
5  title('3D histogram')
```

In `hist3d.sce`, random numbers are amplifiled and plotted on a 2D matrix. The random numbers generate the height of the bar. Two histograms, one in the positive quandrant and one in the negative quadrant, are generated. The result is shown in Figure 4-27.

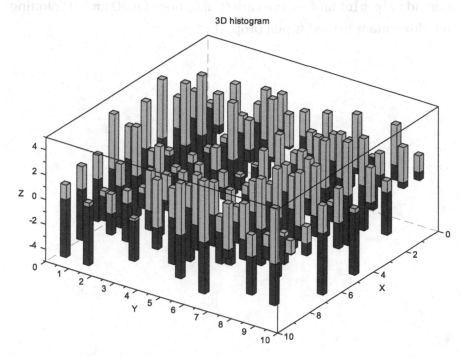

Figure 4-27. *3D histogram*

4.7 Summary

In this chapter, we have presented a variety of built-in functions for 2D and 3D plotting as well as for animations. With these rich features, Scilab provides a nice developing environment for scientists and engineers.

The visualization tools presented in the chapter are comparable to many commercial alternatives. Explaining each and every option for the plots is beyond the scope of this chapter so we have only discussed the most relevant functions. We advise users to study the documentation using the command `help plot` and see the variety of options for 2D and 3D plotting along with options to modify plot properties.

CHAPTER 5

Data Through File Reading and Writing

5.1 Introduction

Using the information in Chapter 4, you can now formulate physical problems in terms of numerical computations and solve them on digital computers. This process calls for the following requirements:

- Data should be in digital form (a digital file).

- Computer programs should be able to read the file and make arrays without errors.

  - If errors have been made, a mechanism to check for those errors and give a warning to the user should be in place.

  - Error correction will be an added feature in this case.

- Data should be stored as an array as per the data type and should be displayed on demand in proper format.

- Array operations on data will result in memory usage in terms of reading and writing data on disks. This should be facilitated by the system. Users should be able to check the status of memory as and when required.

© Sandeep Nagar 2017
S. Nagar, *Introduction to Scilab*, https://doi.org/10.1007/978-1-4842-3192-0_5

- Post-processing tasks include visualizing data in various formats: as a printout from a printer, as a graph on a terminal or printer/plotter, and so on.

- If a report for a particular experiment having input parameters, processing data, and output as a file or graph can be generated, the task is made easier for the user.

Scilab has some features for each of these requirements. In this chapter, we will discuss some of the features that are related to handling files both for inputting and outputting data in various formats.

5.2 File Operations

File operations constitute an important part of solving computations. It is important to note that the file system is OS (Operating System) dependent. Since Unix/Linux has been the popular operating system for scientific and engineering work, Scilab supports Unix/Linux commands for file management. When Scilab is installed on Windows, you use the same commands for dealing with files as you would with Linux since you always works in a Scilab micro-environment. The codes in the following sections have been tested on Windows 8, MacOSX 10.12, and Ubuntu 16.04 systems.

5.2.1 Users

A computing system is accessed by different users using an operating system. Each user defines and then works in a workspace to avoid damaging another user's files. After login, a user's workspace becomes active for that user. Scilab does not provide the facility of login for the software, but when users log into their OS, their workspace for Scilab is activated as well. (A public computer does not provides this facility.)

A workspace is made up of various files and folders. Files store data, and folders/directories are just special files that store links to other files. Hence, multiple files can be stored in one folder.

There are many operations that you can do with a file object. Reading a file involves inputting the data in the file to a desired location. When this location is the computer monitor and you have a text-visualization tool, you can see the data as characters or symbols. When this location is a program to accept the data in the file, it is read by the program and fed to the desired function. Data come in a variety of formats if format interchanging is required as per the requirement of input and output ends. Some files (Scilab's `.sci` and `.sce` files in the present case) need to be *executed* (that is, run the program).

For all of the purposes mentioned, files are opened before an operation and must be closed after an operation. The status of a file (whether it's open or closed) is shown by a flag (usually an integer number).

Some files are essential for the OS to define the behavior of a system. As a result, these files should not be altered. Unwanted alterations are prevented by giving permissions to users. "Reading" and "writing" a file are restricted by permissions. The "administrator" (fondly called `admin`) is also called the "super user" and has all the privileges and permissions to edit any file/folder. You need to understand the defined user type on a particular computer system and then issue commands accordingly. If you are not permitted to access certain folders and input data is placed inside those files/folders, then, unless you ask the `admin` to change the permission parameters, you will always get an error.

On a public computer, `admin` privileges are usually not given. This restricts `read`, `write`, and `execute` operations. When users are on a personal computer, they usually declare themselves the super user and, thus, their account has all the required privileges. On a computer shared by many users, the `admin` assigns privileges to read, write, and execute selectively. Users must seek the appropriate permission for a desired operation.

5.2.2 File Path

Directories/folders can contain subdirectories/subfolders and files.
This behavior can go to any level if this process is not restricted by the
administrator. The pwd command stands for *print working directory*. On
a Scilab terminal, typing pwd displays the path of the present working
directory as shown in the following example:

```
1  -->pwd
2  ans =
3  /Users/<userName>
```

Under the user, the /Users directory contains another directory named
by the username (shown by <userName>) given to the user while installing
the operating system. This is the primary workshop assigned to a Scilab
session. Unless the user changes the directory, the user performs all tasks
in this workspace. When pwd is typed in the terminal, a variable named ans
stores the data (file path). As illustrated before, ans stores the last output at
Scilab REPL. A variable (name chosen by the user) can be assigned to store
the file name as a string.

A file/folder is accessed by writing the file path on the terminal. Let's
do a small exercise to understand this process. To make a new directory,
use mkdir 'name' as shown the following code:

```
1  ->mkdir scilab // make new directory called 'scilab'
2   ans =
3  1.
4  -->cd scilab /// change present directory to the directory
   name 'scilab'
5  ans =
6  /Users/sandeepnagar/scilab
7  -->mkdir test // make a directory named 'test'
8  ans =
9  1.
```

```
10  -->cd test // change present directory to the directory
    name 'test'
11  ans =
12  /Users/sandeepnagar/scilab/test
13  -->ls // List the contents of present directory i.e 'test'
14  ans =
15  []
16  -->mkdir new // make a directory here named 'new'
17  ans =
18  1.
19  -->ls // List the contents of present directory i.e 'test'
20  ans =
21  new
22  -->rmdir new // remove the directory named 'new'
23  ans =
24  1.
25  -->ls // List the contents of present directory i.e 'test'
26  ans =
27  []
28  -->isdir('/Users/sandeepnagar/scilab/test')
    // Checking if a directory 'test' exists
29  ans =
30  T
```

At line number 1, mkdir scilab makes a directory named 'scilab'.
To see the contents of the present directory, we can use the command ls,
which stands for *list*. To change the directory, we can use the command cd
"file path". We will work in this directory for the rest of the exercises in
this book.

We can create a directory named test in the directory named scilab.
We can see the list of its contents by using the command ls. When the

directory is empty, ans=[ ] is displayed. If we now create a new directory named new within the subdirectory test by issuing the command mkdir new and then check for the list of contents, the result is shown as ans=new. To remove a directory, we can use the command rmdir. So, by executing rmdir new, we can see that the user has removed the directory named new.

The command isdir checks whether the input argument (path) directs to a directory. When a directory is found at the input path, the output is the boolean variable T (True); otherwise, the output is F (False).

5.2.3 Creating Files and Saving Them

The save and load commands enable us to write and read data to the memory:

```
1   -->a = rand(3,3)
2   a =
3   0.2113249    0.3303271    0.8497452
4   0.7560439    0.6653811    0.6857310
5   0.0002211    0.6283918    0.8782165
6   -->save('rand_matrix.dat','a')
7   -->ls
8   ans =
9   ! rand_matrix.dat    !
10  -->load('rand_matrix.dat','a')
11  -->a =
12  0.2113249    0.3303271    0.8497452
13  0.7560439    0.6653811    0.6857310
14  0.0002211    0.6283918    0.8782165
```

At line number 1, a variable named a, which stores a random value 3 × 3 matrix, is created first. At line number 2, this data is stored as a .dat file (data file) named rand_matrix.dat, which passes the variable name as the argument. When required, this file can be loaded in the workspace using the load command. Please note that the data don't always have to be comprised of numbers.

The data can be anything that a digital computer can handle, including pictures, videos, strings, and characters, just to name a few.

Multiple variables can be stored in the same file by passing the name of variables at the time of saving:

```
1   -->a = rand(2,2)
2   a =
3   0.0683740      0.6623569
4   0.5608486      0.7263507
5   -->b = rand(3,3)
6   b =
7   0.1985144      0.2312237      0.6525135
8   0.5442573      0.2164633      0.3076091
9   0.2320748      0.8833888      0.9329616
10  -->c = rand(3,2)
11  c =
12  0.2146008      0.2922267
13  0.312642       0.5664249
14  0.3616361      0.4826472
15  -->save('rand_matrix1.dat','a','b','c')
16  -->ls
17  ans =
18  !rand_matrix1.dat    !
19  !                    !
20  !rand_matrix.dat     !
21  -->load('rand_matrix1.dat','a','b','c')
22  -->a
```

```
23  a =
24  0.0683740     0.6623569
25  0.5608486     0.7263507
26  -->b
27  b =
28  0.1985144     0.2312237     0.6525135
29  0.5442573     0.216463      0.3076091
30  0.2320748     0.8833888     0.9329616
31  -->c
32  c =
33  0.2146008     0.2922267
34  0.312642      0.5664249
35  0.3616361     0.4826472
```

help("save") and help("load") give very useful instructions about using these commands.

The load() function follows the same logic as the save() function. Data can be unzipped and loaded from a particular formatted file as an array. An array, thus populated, can be used for computation, and resultant files can be made using the save() function again (if required). Elaborate computations require this procedure to be repeated successively many times. Thus, the functions have been optimized to locate and load required data in a short time.

5.2.3.1 Opening and Closing Files

To read and write data files, they must be opened and defined as readable and/or writeable. The command mopen() is used to create a nonexistent file or to open an existing file. For example, the following code shows how to create a file named test.sce in write mode. A variable named file_handle is used to store the event. Looking at the list of contents of the folder, you can see that a file by this name is indeed created. Similar to the mopen() command, the mclose() command is used to close the file:

```
1  --->file_handle = mopen('test.sce','w') // creates the file
   in 'write' mode
2  file_handle =
3  1.
4  --->ls
5  ans =
6  test.sce
7  --->mclose(file_handle) // close the file and an integer 0
   is returned signifying successful closing of file
8  ans =
9  0.
```

The mode parameter controls the access type for the file that can have one of the values found in Table 5-1.

Table 5-1. *Different File Modes*

r	opens for reading (default). The file must exist; otherwise, it fails.
w	opens for writing. If the file exists, its contents are destroyed.
a	opens for appending. It creates the file if it does not exist.
r+	opens for both reading and writing. The file must exist; otherwise, it fails.
w+	opens for both reading and writing. If the file exists, its contents are destroyed.
a+	opens for both reading and appending. It creates the file if it does not exist.

In addition, the characters in Table 5-2 can be used to specify the type of file.

Table 5-2. *Different File Types*

t	text file
b	binary file (default)

The default access mode is `'rb'` (binary file reading).

5.2.3.2 csvread() and csvwrite()

The functions csvRead() and csvWrite() are used to read data from the .csv file, which stands for *comma separated values*:

```
1  -->a = rand(4,4);
2  -->csvWrite(a,'csv file data');
3  -->b = csvRead('csv file data');
4  -->a
5  a =
6
7  column 1 to 2
8
9  0.3321719    0.2693125
10  0.5935095    0.6325745
11  0.5015342    0.4051954
12  0.4368588    0.9184708
13
14  column 3 to 4
15
16  0.0437334    0.2806498
17  0.4818509    0.1280058
```

```
18   0.2639556      0.7783129
19   0.4148104      0.2119030
20
21   -->b
22   b =
23
24   column 1 to 2
25
26   0.3321719      0.2693125
27   0.5935095      0.6325745
28   0.5015342      0.4051954
29   0.4368588      0.9184708
30
31   column 3 to 4
32
33   0.0437334      0.2806498
34   0.4818509      0.1280058
35   0.2639556      0.7783129
36   0.4148104      0.2119030
```

The functions to write and read a .csv file have a number of options that can be understood by using the command help("csvRead") and help("csvWrite"). Users are encouraged to explore the options since .csv files are used abundantly in the domain of data analytics. Using and manipulating them is an important skill of a data scientist.

Note   A number of other functions to read and write files exists, but we have focused only on some of the most commonly used ones. Documentation can be accessed to explore other specialized functions, if required.

5.3 Summary

In this chapter, we have illustrated various functions enabling reading and writing permissions, as well as taking data to and from a file. These functions comprise an essential part of a numerical computation exercise. The data can be generated in the form of files using a software package or hardware (an instrument). Scilab does not care about its origin. It treats data by their type and file type. Judging an appropriate function to operate using files has to be done by the user as the situation requires. File operations provide faculties to trim the data so that only the useful part of the data is fed as an array. Further trimming can be performed by slicing operations. With the art of handling files, you can confidently proceed to handling sophisticated numerical computations.

CHAPTER 6

Functions and Loops

6.1 Introduction

When a particular numerical task needs to be "repeated" over different data points, digital computers become a useful tool since they can perform this action with greater speeds than humans. Loops perform exactly this task. Using a condition to check the start and termination rules, users can perform repetitive parts of a process as desired. Different programming languages and environments have different rules for defining loops. Scilab provides a simple way to define and run loops. In addition to loops, functions also define an important part of modern programming architecture. A big program may require a set of instructions to be called at different times. Hence, this set of instructions can be defined as a subprogram, which can be requested to perform the computation at a desired time. In this way, a complicated task can be divided into many small parts. This architecture of programming is called *modular programming*. It is the most popular way of programming since it is quite logical, good at visualizing the problem, and easy to debug. The most popular way of defining these small sets of instructions is to define them as functions. Together, functions and loops break a numerical computation problem into a series of simpler problems that can be accessed as required. In this chapter, we will discuss both of these concepts in detail.

© Sandeep Nagar 2017
S. Nagar, *Introduction to Scilab*, https://doi.org/10.1007/978-1-4842-3192-0_6

6.2 Loops

Loops form an essential part of an algorithm since they perform the tasks that computers perform best: doing repetitive actions in a very fast manner. Loops can come in many flavors such as for loop, which repeats certain tasks over a list of variable values; while loop, which checks a logical condition before executing a certain task; and if-then-else loop, which checks a condition and directs the flow of an algorithm. The loop you choose depends on the problem at hand.

A variety of functions and their usage are described in the following sections. Judging their usage critically becomes supremely important because the looping part of an algorithm consumes most of the execution time.

6.2.1 while

while loop defines a logical condition and, until it is satisfied, it runs a block of code. The syntax for while loop is the following:

```
1   while condition
2   BODY
3   endwhile
```

Here, the keyword while initiates the execution of a while loop. The condition is a logical condition whose answer can be 'true' (1) or 'false' (0). The BODY encompasses the string of commands that is executed until the condition holds true, as shown in Listing 6-1.

Listing 6-1. while.sce

```
1   i = 1
2   while i<20
3       disp(i);
4       i = i*2;
5   end
```

The result is shown as follows:

```
1  -->exec('/Users/sandeepnagar/.../while.sce', -1)
2  1.
3  2.
4  4.
5  8.
6  16
```

The code while.sce first initializes the variable i to the numerical value 1. It then checks the condition $i < 20$. This is true at the first step when $i = 1$. Consequently, it enters the loop and executes the command disp(i). (The numerical value 1 is printed on the terminal.) Then the next line is executed ($i = i * 2$), which makes the new value of $i = 2$. At the end of the loop, the control is further taken by the condition statement ($i < 20$). Until this holds true, the loop runs and, hence, 1, 2, 4, 8, 16 are printed. When $i = 16$ and the statement $i = i * 2$ is executed, the new value of i becomes 32. Now the condition is not satisfied and the loop is terminated.

6.2.2 Infinite Loops

Some loops can run infinitely so they are called *infinite loops*. Try the code while.sce by initializing $i = 0$ instead of $i = 1$. In this case, the value of i will always be 0 inside the loop and the condition $i < 20$ will always be true. Hence, the code will run forever if it is not interrupted. Infinite loops can be interrupted by the Ctrl+C key combination on an an ASCII keyboard.

The onus of avoiding infinite loops lies with the user. Scilab will simply execute the statements mindlessly without showing a warning or error message. Technically, infinite loops are not programming errors as they are syntactically correct. In fact, infinite loops can be used if you need to generate an infinite stream of data or if you need to execute a Scilab program infinitely. For example, suppose a Scilab program reports the status of a remote wind turbine. This code needs to run infinitely unless interrupted. In another scenario, suppose an application requires a stream of random numbers. In this case, a Scilab code can be written to generate an infinite sequence of random numbers.

6.2.3 for

for loop is used to perform computation on a list of known values. The syntax of for loop is the following:

```
1  for variable = vector
2      BODY
3  end
```

The keyword for declares the starting of the loop where a variable takes the values stored in a vector. Then a body of code (represented by BODY) is executed. The keyword end declares the end of for loop. This is explained in Listing 6-2.

Listing 6-2. for.sce

```
1  for i = 1:10
2      square_root = sqrt(i);
3      disp(square_root)
4  end
5
6  disp("Program finished")
```

Executing for1.m yields:

```
1  --->exec('/Users/sandeepnag.../for.sce', -1)
2  1.
3  1.4142136
4  1.7320508
5  2.
```

```
 6   2.236068
 7   2.4494897
 8   2.6457513
 9   2.8284271
10   3.
11   3.1622777
12   Program finished
```

The for statement creates a vector of numerals from 1 to 10 and stores it in a variable named *i*. Each member of this vector is fed to the body of the loop. A variable named square_root stores the square root of the value stored in the array. It is then printed on a Scilab terminal using the command disp(square_root). When these two commands are finished, the next member of the vector is picked and the same is repeated. This is continued until the last member of the vector is stored in the variable named *i* (*i* = 10).

6.2.4 if-elseif-else

In situations where a number of conditions needs to be checked at different points of time, if-elseif-else loop works well. The syntax for this loop is given by the following:

```
1   if condition1
2   BODY1
3   elseif condition2
4   BODY2
5   else
6   BODY3
7   endif
```

At line 1, a condition is defined. If this condition is satisfied, then line 2 is executed or else line 3 is executed. Hence, BODY1 and BODY2 are the blocks of codes that are executed by checking for different set of conditions, and BODY3 set of codes is executed in the case when none of the condition is executed. (See Listing 6-3.)

Listing 6-3. ifelse.sce

```
1  i = rand(1,1);
2  if i>0.5 then
3  disp(i);
4  disp("True");
5  else
6  disp(i);
7  disp("False");
8  end
```

Running the code yields the following:

```
1  --->exec('/Users/sandeepnagar/.../ ifelse.sce', -1)
2  0.5376230
3  True
4  --->exec('/Users/sandeepnagar/.../ifelse.sce', -1)
5  0.1199926
6  False
7  --->exec('/Users/sandeepnagar/.../ifelse.sce', -1)
8  0.2256303
9  False
```

Whenever the value of a random number is more than 0.5, True is printed; False is printed otherwise. (See Listing 6-4.)

Listing 6-4. ifelseif.sce

```
1   i = rand(1,1);
2   if i>0.5 then
3   disp(i);
4   disp("Value is larger then 0.5");
5   elseif i>0.3 then
6   disp(i);
7   disp("value is larger than 0.5 and 0.3");
8   else
9   .disp(i)
10  disp("value is smaller than 0.5")
11  end
```

When executing the code ifelseif.sce, we obtain the following results on a Scilab terminal:

```
1   -3->exec('/Users/sandeepnagar/.../ifelseif.sce', -1)
2   0.0485566
3   value is smaller than 0.5
4   -3->exec('/Users/sandeepnagar/.../ifelseif.sce', -1)
5   0.6723950
6   Value is larger then 0.5
7   -3->exec('/Users/sandeepnagar/.../ifelseif.sce', -1)
8   0.2017173
9   value is smaller than 0.5
10  -3->exec('/Users/sandeepnagar/.../ifelseif.sce', -1)
11  0.3911574
12  value is larger than 0.5 and
13  0.3
```

6.3 Functions

A function is a set of codes that can be called when required. As a result, it can be defined separately either its own file or within the body of the program. A script file is similar in nature. A script file stores a sequence of commands to be executed. It seems that a function and a script have a similar nature, but, unlike MATLAB and Octave, Scilab provides separate kinds of files for each one of them. This is based on the nature of their behavior with the core Scilab program.

Whereas a script file (with extension .sce) is an executable file, a function file (with extension .sci) stores a set of instructions. The function file behaves like a black box where input is fed and output is obtained. On the other hand, a script file changes its behavior as per input values. Whatever input data a script file accesses is taken from the Scilab workspace. Output data from a script file is put into the Scilab workspace. The semantics of input data, local variables, are visible only within the function.

The definition of a function follows this syntax:

```
1   function [o1,o2,...] = function_name (i1,i2,...)
2   statement_1
3   statement_2
4   ...
5   statement_n
6   endfunction
```

Here the function keyword defines the object types as function. Then a set of variables is defined that this function is expected to return (o1,o2,... signifying output1, output2,...). Next comes an = operator and then the name of the function. In the previous case, it is function_name. A function take inputs (i1,i2,... signifying input1, input2,...) to produce an output according to calculation defined in its body. Then comes the main body of the function where commands for executing the purpose of the function are mentioned. The last statement, endfunction, signifies the end of the function.

For example, we can write a function to find $x^2 - y^2$ and assign it to variable name z, as shown in Listing 6-5.

Listing 6-5. fn1.sci

```
1  function y = fn1(a,b)
2  y = a^2-b^2;
3  endfunction
```

Notice that the extension of this code is .sci. This file must first be loaded in a Scilab workspace. We need to provide the full path of the file to the built-in function exec()first and then use the function by providing its name with input arguments:

```
1  --->exec('/Users/sandeepnagar/.../fn1.sci', -1)
2  --->fn1(2,3)
3  ans =
4  -5.
```

It is good practice to define the program as a group of *function files* and call them in the master program stored as a *script file*. This modular approach makes it easy to experiment with the idea and also makes it easier to debug and test the code. A function can return more than two values as well, as shown in Listing 6-6.

Listing 6-6. fn2.sci

```
1  function[y1,y2,y3] = fn2(x,y)
2  y1 = x - y;
3  y2 = x + y;
4  y3 = y - x;
5  endfunction
```

This gives the following result:

```
1  --->exec('/Users/sandeepnagar/.../fn2.sci', -1)
2  --->[a,b,c] = fn2(2,3)
3  c =
4  1.
5  b =
6  5.
7  a =
8  -1.
```

Functions can incorporate loops to regulate the repetitive tasks inside the program. For example, the factorial of a number can be calculated using a function given in Listing 6-7.

Listing 6-7. factorial1.sci

```
1  function result = factorial1(n)
2    if(n == 0)
3        result = 1;
4        return;
5    else
6        result = prod(1:n);
7    end
8  endfunction
```

A function named `factorial1`, which takes a number `n` as an argument, calculates the product of the number with all its successive numbers. When called from a Scilab command line, the function yields the following result:

```
1  --->exec('/Users/sandeepnagar/.../factorial1.sci', -1)
2  --->factorial1(10)
3  ans =
4  3628800.
```

```
5  --->factorial1(10e5)
6  ans =
7  Inf
```

6.3.1 Inline Functions

An inline function is a short function that can be defined without having to use the function skeleton discussed previously. This is useful only when the body of the function is short:

```
1  --->deff('[x] = mult(y,z)','x=y*z')
2  --->mult(2,3)
3  ans =
4  6
```

The built-in function `deff()` is used to define another built-in function. The first argument of `deff` defines the output variable (x in this case), function name (`mult`), and input variables (y and z). The second argument defines the body of the function (x=y*z). This kind of function is defined in a `.sce file`, just like a command, and thus does not need a separate loading action. Once defined, functions can be called by their function name along with input parameters.

6.4 Summary

Defining functions is the key to modular programming. Scilab presents an elegant way to define and use functions both inline and in separate files. When combined with the ability to write functions inside a loop, complex problems can be implemented in a few lines of codes. It requires an artistic attitude while designing an algorithm where functions and loops are the paintbrush to devise an elegant solution to a given numerical problem.

CHAPTER 7

Numerical Computing Formalism

7.1 Introduction

Numerical computation enables us to compute solutions for physical problems, provided we can frame them into a proper format. This process requires certain considerations. First and foremost is the understanding of approximate solutions. For example, if we digitize continuous functions, then we are going to introduce certain errors due to the sampling at a finite frequency. Hence, a very accurate result would require a very fast sampling rate. In cases when a large data set needs to be computed, it becomes computationally an intensive and time-consuming task. Users need to understand that the numerical solutions are an approximation at best when compared to analytical solutions. The onus of finding their physical meaning and significance lies with user. The art of discarding solutions that do not have a meaning for real-world scenarios is a skill that a scientist or engineer develops over the years. Also, a computational device is just as intelligent as its operator. The law of GIGO (garbage in, garbage out) is followed very strictly in this domain.

In this chapter, we will discuss some of the important steps involved in solving a physical problem using numerical computation. Defining a

© Sandeep Nagar 2017
S. Nagar, *Introduction to Scilab*, https://doi.org/10.1007/978-1-4842-3192-0_7

problem in proper terms is just the first step. Making the right model and then using the right method to solve (solver) the problem distinguishes the experienced scientist/engineer from the naïve one.

7.2 Physical Problems

Everything in our physical world is governed by physical laws. Thanks to scientists who toiled under difficult circumstances and came up with fine solutions to phenomena happening around us, we obtained mathematical theories for physical laws. To test these mathematical formalisms of physical laws, we use computational techniques. Analytical computation involves the use of symbols. Numerical computation, instead, involves the use of computers, which follow binary logic. Both approaches have their limitations. Even though analytical solutions are very accurate, they are difficult to derive, especially if a predefined framework is absent. On the other hand, numerical solutions are always an approximation of real values. In any case, if the computation yields the same results as that of a real experiment, the two results validate each other. Numerical simulations can remove the need of doing an experiment altogether provided we have a well-tested mathematical formalism. For example, nuclear powers of our times don't test nuclear bombs "for real" any more. The data about nuclear explosion, which were obtained during real nuclear explosions, enabled scientists to model these physical systems quite accurately, thus eliminating the need for real testing.

In addition to applications like simulating a real experiment, modeling physical problems is a good educational exercise. While modeling, hands-on exercises enable students to explore a subject in depth and give proper meaning to the topic being investigated. Solving numerical problems and the visualization of results make the learning permanent and also ignites research regarding flaws in mathematical theory, which ultimately leads to new discoveries.

7.3 Defining a Model

Modeling means writing equations for a physical system. As the name suggests, an equation is about equating two sides. An equation is written using an = sign where terms on the left-hand side equal terms on the right-hand side. The terms on either side of equations can be numbers or expressions. For example:

$$3x + 4y + 9z = 10$$

This is an equation having one term, $3x + 4y + 9z$, on the left-hand side (LHS) and one term, 10, on the right-hand side (RHS). Please note that whereas LHS is an algebraic expression, RHS is a number.

Expressions are written using functions, which are simply relations between two domains. The function $f(x) = y$ is a relation between y and x using rules of algebra. Mathematics has a rich library of functions that we can use to make expressions. The function we choose depends on the problem. Some functions describe certain situations better than others. For example, oscillatory behavior can be described in a reasonable manner using trigonometric functions like $sin(x)$ and $cos(x)$. Objects moving in straight lines can be described well using linear equations like $y = mx + c$ where x is their present position, m is the constant rate of change of x with respect to y, and c is the offset position. Objects moving in a curved fashion can be described by various nonlinear functions (where the power of the dependent variable, like x in the previous example, is not 1).

In real life, we can have situations that are a mixture of these scenarios. For example, an object can oscillate and move in curved fashion at the same time. In this case, we would write an expression using a mixture of functions or find new functions that could explain the behavior of the object. Verification of these choices of functions is done by finding solutions to equations describing the behavior and matching it with observations made about an object. If they match perfectly, we obtain perfect solutions. In most cases, an exact solution might be difficult to obtain. In these cases, we get an "approximate" solution. If the errors involved while obtaining an approximate solution are within tolerance limits, the models can be acceptable.

As previously discussed, physical situations can be analytically solved by writing mathematical expressions in terms of functions involving dependent variables. The simplest problems have simple functions between dependent variables with a single equation. There can be situations where multiple equations are needed to explain a physical behavior. In cases of multiple equations being solved, the theory of matrices comes in handy.

Suppose Equations 7.1 and 7.2 define the physical behavior of a system:

$$-x + 3y = 4 \tag{7.1}$$

$$2x - 4y = -3 \tag{7.2}$$

Then this system of two equations can be represented by a matrix equation, as follows:

$$\begin{bmatrix} -1 & 3 \\ 2 & -4 \end{bmatrix} + \begin{bmatrix} x \\ y \end{bmatrix} = \begin{bmatrix} 4 \\ 3 \end{bmatrix}$$

Now, using matrix algebra, the values of variables x and y can be found such that they satisfy the equations. Those values are called *roots* of these equations. These roots are the point in 2D space (because we have two dependent variables) where the system will find stability for that physical problem. In this way, we can predict the behavior of a system without actually doing an experiment.

The mathematical concept of differentiation and integration becomes very important when we need to work with a dynamic system. When the system is constantly changing the values of dependent variables to produce a scenario, it becomes important to know the rate of change of these variables. When these variables are independent of each other, we use simple derivatives to define their rate of change. When they are not independent of each other, we use partial derivatives for the same.

For example, Newton's second law of motion indicates that the rate of change of velocity of an object is directly proportional to the force applied on it. Equation 7.3 demonstrates this concept mathematically:

$$F \propto \frac{dv}{dx}$$

(7.3)

The proportionality is turned into equality by substituting for a constant of multiplication m such that

$$F = m \times \frac{dy}{dx}$$

(7.4)

If we know the values or expressions for F, this equation can be solved analytically and solutions can be found to this equation. But, in some cases, the analytical solution may be too difficult to obtain. In such cases, we digitize the system and find a numerical solution.

There are many methods to digitize and numerically solve a given function. Programs to implement a particular method to solve a function numerically are called solvers. Many solvers exist to solve a function. The choice of solver is critical to successfully obtain a solution. For example, Equation 7.4 is a differential equation. It is a first-order ordinary differential equation. A number of solvers exist to solve it including Euler and Runge-Kutta. The choice of a particular solver depends on the accuracy of its solution, the time taken for obtaining a solution, and the amount of memory used during the process. The latter is important where memory is not a freely expendable commodity like microcomputers with limited memory storage (for example, RaspberryPi).

The advantage of using Scilab to perform a numerical computation lies in the fact that it has a very rich library of functions to perform various tasks required. The predefined functions have been optimized for speed and accuracy (in some cases, accuracy can be predefined). This enables the user to rapidly prototype the problem instead of concentrating on writing functions to do basic tasks and optimizing them for speed, accuracy, and memory usage.

7.4 Scilab Packages

A number of packages exists to perform numerical computations in a particular scientific domain. Scilab uses the Atoms package manager for managing (installing, deleting, and updating) a Scilab package. The web site `https://wiki.scilab.org/ATOMS` describes its usage.

Clicking the menu item `Applications -> Module Manager ATOMS` opens up the Atoms (AuTomatic mOdules Management for Scilab) package manager. It lists a variety of specialized modules for Scilab. It is frequently updated with new developments. You can select a package by clicking its name and install it using the Install button. In a similar way, you can uninstall the package. You can also perform these actions using a Scilab command line.

7.4.1 Searching a Package

If you know the name of the package, you can search for the same using the following command:

```
1  -->atomsSearch("coselica")
2  ans =
3  !coselica Standard Open Modelica Blocks   !
```

```
4  -->atomsSearch("arduino")
5  ans =
6  !arduino Arduino Communication through Serial   !
7  -->atomsSearch("raspberrypi")
8  ans =
9  []
```

Whereas the modules named coselica and arduino could be found, the module named raspberrypi returned an empty string. (In other words, it could not be found.)

7.4.2 Installing a Package

The commands to install a module use the built-in function atomInstall() as follows:

```
1  -->atomsInstall("scidemo")
2  ans =
3
4  !scidemo 0.2.2  allusers     SCI/contrib/scidemo
   /0.2.2  I   !
5  !                                                !
6  !apifun  0.4.2  allusers     SCI/contrib/apifun
   /0.4.2   A   !
```

The returned argument shows the version number of the module as well as the path at which it is stored. The module is loaded during the next startup run of Scilab automatically. To load a module manually, you can use the command atomsLoad("scidemo").

7.4.3 Removing a Package

The built-in function atomsRemove("scidemo") can be used to remove the package arduino:

```
1  ——>atomsRemove("scidemo")
2  ans =
3
4  !scidemo 0.2.2  allusers    SCI/contrib/scidemo
   /0.2.2   I    !
5  !                                                              !
6  !apifun  0.4.2  allusers    SCI/contrib/apifun
   /0.4.2    A    !
```

7.4.4 Listing Packages

The built-in function atomsList() can be issued at the command line to obtain a detailed list of packages sorted in alphabetical order. For example, the following output is produced at the time of writing:

```
1  ——>atomsList
2  apifun — Check input arguments in macros
3  assert — A collection of predicate functions
4  condnb — Evaluates the condition number of functions.
5  coselica — Standard Open Modelica Blocks
6  CPGE — CPGE dedicated Xcos blocks
7  dbldbl — Double—Double floating point numbers
8  diffcode — Automatic differentiation
9  dispmat — Display matrices graphically
10 Dynpeak — Scilab toolbox for the detection of pulses in
   hormonal signals
```

11 FACT — a toolbox for chemometric applications , e.g.
    regressions , discriminant analysis , multiway analysis

12 floatingpoint — Functions to manage floating point numbers

13 guibuilder — A Graphic User Interface Builder

14 helptbx — Update the help of a module automatically .

15 iodelay — manipulation and frequency analysis of linear
    dynamical systems with input or output delays

16 makematrix — A collection of test matrices .

17 Mathieu — Solve Mathieu equations , calculate Mathieu
    functions

18 mingw — Dynamic link with MinGW for Scilab on Windows

19 neuralnetwork — This is a Scilab Neural Network Module
    which covers supervised and unsupervised training
    algorithms

20 number — Integers algorithms

21 optkelley — Scilab software for Iterative Methods for
    Optimization

22 removed — All pages of functions removed from Scilab 4.1.2
    to 6.0 gathered in a single place

23 scibench — A collection of benchmarks

24 scicv — Interface to the Computer Vision library OpenCV

25 scidemo — A collection of demonstrations

26 serial — A toolbox for communication over a Serial Port in
    Scilab

27 specfun — A collection of special functions

28 stixbox — Statistics toolbox

29 uman — User MANual in console + online + browser , easy
    languages switch , related bugs , comb mailing lists ...

30 uncprb — Provide 35 unconstrained optimization problems

7.5 XCOS

One of the most attractive features of Scilab (which is absent in both MATLAB and Octave) is an XCOS graphical programming environment. It is similar to MATLAB's graphical programming environment called Simulink. A detailed discussion of XCOS is the topic of Chapter 8.

7.6 Summary

Almost all branches of science and engineering require the performance of numerical computations. Scilab is one of the alternatives for doing such computations. Scilab has a rich library of optimized functions for general computation that is growing day-by-day as developers all around the world are contributing to the effort. Also, it has a variety of packages for performing specialized jobs. This makes Scilab an ideal choice for prototyping a numerical computation problem efficiently.

CHAPTER 8

XCOS

8.1 Introduction

Historically, MATLAB has had one feature that makes it stand out among others when it comes to ease of teaching computing—Simulink. Simulink is not part of basic MATLAB, so it must be purchased separately. With Simulink, you can make a program by visually connecting block of codes. Scilab provides an equivalent to XCOS. XCOS is a toolbox for the modeling and simulation of dynamic (continuous and discrete) systems. Although its main purpose is to simulate dynamic systems, XCOS can also be used for signal generation, data visualization, and simple algebraic operations. While simulating systems that deal with interconnected continuous-time and discrete-time components, XCOS will fit perfectly for modeling and simulation.

XCOS is one of the most powerful tools from Scilab for new users. Because it is a GUI-based programming environment, Scilab just requires users to connect blocks to write a code. These blocks of code can be dragged and connected as the users wish. The flow of the information within a code is defined by the directions and usage of connectors between the code block. The blocks are actually visual representations to generalized code whose inputs and outputs can be changed as per usage.

XCOS provides a *modular* approach for complex system modeling, using a block diagram visual editor. The modular approach makes the

© Sandeep Nagar 2017
S. Nagar, *Introduction to Scilab*, https://doi.org/10.1007/978-1-4842-3192-0_8

activity of developing a simulation quite simple since users can just replace a module with another one to test the system under different scenarios. XCOS models are compiled and simulated in a single run. The resulting mathematical equations are integrated by a numerical solver with configurable parameters.

8.2 Installing XCOS

XCOS comes pre-installed with the latest version of Scilab, but if you are using older versions, you can easily install it using the Atoms package manager, just like any other package. The graphical way to install it is to click Applications in the menu bar and then click the module manager Atoms. Next, you find XCOS and click Install. You need an active Internet connection for this activity and the installaltion time depends on the time taken to download the package and then the speed of computing resources.

An offline installation would require you to download the XCOS package as a .zip file at a particular location and then issue the following command at at a Scilab command terminal:

```
1  -->atomsInstall("download_path\file_name_zip_")
```

Here you substitute a relevant download path of the .zip file that was downloaded.

8.3 Launching XCOS

XCOS can be launched using command lines as well as clicking an icon in the toolbar. It is important to note that you need a graphics terminal for this purpose. Most modern computers running the latest operating systems have graphics capabilities The graphic visualizations of XCOS depends on the native operating system's graphics configurations.

Most often, a new user doesn't need to play with graphics configurations to work with XCOS, but if problems persist, consulting the system administrator is a good option.

8.3.1 Using a Command Line

At the Scilab command line, you can simply type the command xcos to launch the XCOS environment. A palette browser and a simulation editor open up for th euser to edit. A sample screenshot of a computer screen is shown in Figure 8-1.

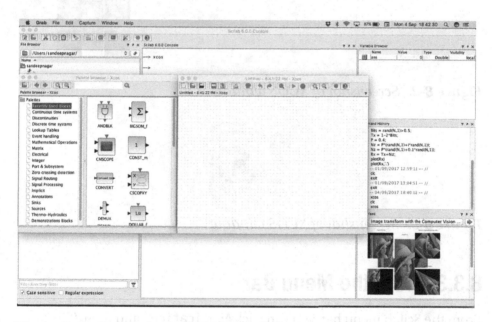

Figure 8-1. *Screenshot of XCOS window*

8.3.2 Using a Graphical Symbol

You can also open XCOS by clicking a symbol in the main toolbar of Scilab, as highlighted by an arrow in Figure 8-2. The symbol in the toolbar looks like Figure 8-3.

169

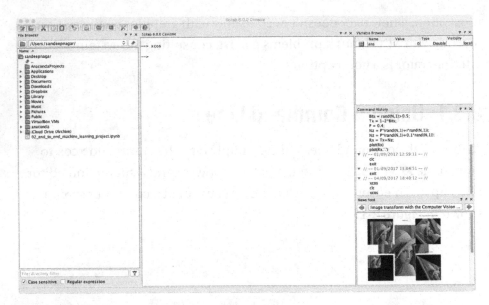

Figure 8-2. *Screenshot of XCOS window*

Figure 8-3. *Screenshot of XCOS window*

8.3.3 From the Menu Bar

From the Scilab menu bar, you can click Applications and then XCOS to
open up an XCOS session.

8.4 XCOS Palettes

The XCOS palette window (as seen in Figure 8-4) allows you to choose
blocks clubbed under a variety of applications.

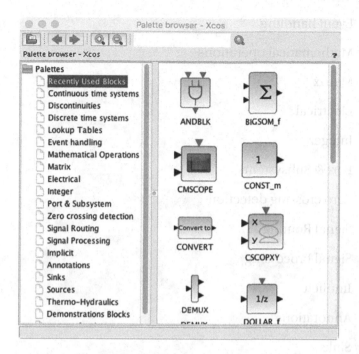

Figure 8-4. *Screenshot of XCOS's palette window*

XCOS palette windows display a library of all available libraries of blocks to be used. They can be dragged to the XCOS editor window (8.5) and dropped there to be connected to each other. The palette browser has two panes. The one on the left presents the list of available predefined palettes (libraries) sections as per different functionalities. (See Figure 8-2.)

- Recently used blocks

  - As you use XCOS, you use some blocks more frequently than others. They are stored here so that you don't have to look for them each time.

- Continuous time systems

- Discontinuities

- Discrete time systems

- Lookup tables

171

- Event handling

- Mathematical operations

- Matrix

- Electrical

- Integer

- Port & Subsystems

- Zero crossing detection

- Signal Routing

- Signal Processing

- Implicit

- Annotations

- Sinks

- Sources

- Thermo-Hydraulics

- Demonstrations Blocks

- User-defined functions

The right pane contains the available blocks for each palette. By clicking another palette in the left pane, a new set of blocks will appear on the right pane.

8.5 XCOS Editor

The editing window is the Xcos workspace for developing new models (diagrams). The XCOS editor window (as seen in Figure 8-5) allows you to design a model by choosing palettes from the palette window, dragging them with a mouse, and releasing them at the editor. You can also add block to the diagram by right-clicking the block in the library and Add to -> "name of the diagram".

Figure 8-5. *Screenshot of XCOS's editor window*

Multiple blocks appear on editor windows and they have input and output ports. These ports can be connected to make a model. Multiple blocks can be connected in a desired way to define a particular simulation. A simulation can then be run by clicking the arrow icon on the toolbar of the editor window. If the simulation is not timed to stop, it will keep running. You can stop it by clicking the Stop button, which is placed next to the Start button. The following buttons are also found in the editor in order from left to right:

- New: To start a new diagram

- Open: To open an existing diagram saved anywhere on computer

- `Open file in Scilab Current Directory`: To open an existing diagram saved in the current working directory

- `Save`: To save a diagram in the computer

- `Save as`: To save a diagram in any other format such as `zcos` (Zipped XCOS file), `xcos` (XCOS file), `xmi` (Eclipse EMF file)

- `Print`: To print the diagram to a printer or as a PDF file

- `Delete`: To delete a block

- `Undo`: To undo the last action performed. This can be done successively.

- `Redo`: To restore the previous state after performing an Undo action. This action can also be done successively.

- `Fit diagram or blocks to view`: Sometimes the area required to make the diagram extends the screen area and scroll bars appear on this window. By using this button, the whole diagram can be made to fit the size of window.

- `Start`: To start a simulation

- `Stop`: To stop a simulation

- `Zoom in`: To zoom in the view of diagram

- `Zoom out`: To zoom out the view of diagram

- `Xcos Demonstrations`: To run some basic demonstrations

- `Help`: To seek help on XCOS commands where you can type queries and then be directed to useful and relevant documentation

Let's start a simple demonstration diagram by clicking the Xcos Demonstrations button. We will get a window similar to the one shown in Figure 8-6.

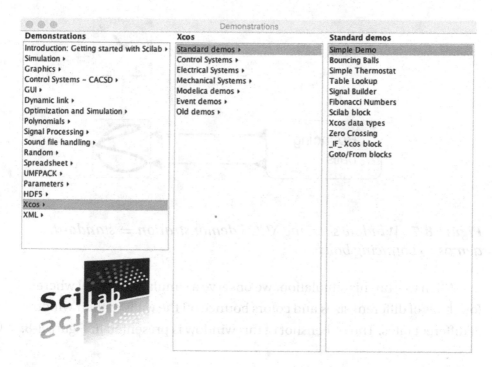

Figure 8-6. *Window showing XCOS demonstration options*

By clicking Standard Demos and then Bouncing Balls, we get a window similar to the one shown in Figure 8-7.

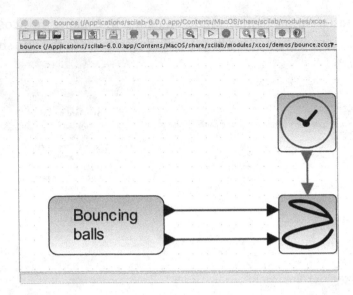

Figure 8-7. *Window showing XCOS demonstration ⟹ standard demos ⟹ bouncing balls*

When we run this simulation, we observe a simulation model where four balls of different sizes and colors bounce off the walls of their window at different rates. The screenshot of this window is presented in Figure 8-8.

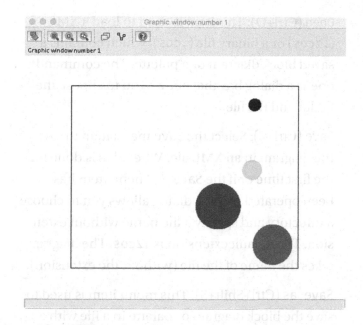

Graphic window number 1

Graphic window number 1

Figure 8-8. Screenshot of window showing bouncing ball animation

8.6 XCOS Menu Bar

In this section, we will describe the menu bar of XCOS so that beginners can navigate their workflow efficiently. The XCOS menu bar exists at the top of the XCOS window. Shortcut keys are displayed by the side of the command and they have been mentioned here as well. Most of the time, these keys use the Control key on the keyboard, which is depicted by Ctrl. When keys must be pressed together, the key combinations are shown by a plus sign in between. For example Ctrl+N means pressing the Control and N keys together.

- File

  - New (Ctrl+N): This menu item opens a new empty Xcos diagram in a new window

- Open (Ctrl+O): This item is used to load a XML (.zcos) or a binary file (.cos), which contains a saved block diagram or a palette. The command opens a dialog box that allows you to choose the folder and the file.

- Save (Ctrl+S): Select the Save menu item to save the diagram in an XML file. When this is done for the first time or if the Save As menu item has not been operated upon, a dialog allows you to choose a directory and specify a file name without extension. The default extension is .zcos. The diagram takes the name of the file (without the extension).

- Save as (Ctrl+Shift+S): This menu item is used to save the block diagram or palette in a file with a new name. A dialog box allows specifying a file name without the extension and a saving folder. The default extension is .zcos. The diagram takes the name of the file (without the extension).

- Export (Ctrl+E): This item is used to export a figure of the current XCOS diagram. The export can be done in the wbmp, gif, html, JPEG, JPG, png, svg, and vml formats.

- Recent Files: This menu item allows quick access to the recent opened files.

- Print (Ctrl+P): This item prints the current diagram onto a printer.

- Close (Ctrl+W): When several diagrams are opened, this item closes the current diagram.

- Quit (Ctrl+Q): The menu item will close XCOS.

- Edit

    - Undo (Ctrl+Z): Select the Undo menu item to undo the last edit operation in case of a mistake.

    - Redo (Ctrl+Y): Select the Redo menu item to redo the last undo edit operation.

    - Cut (Ctrl+X): This item is used to remove the selected objects from the diagram and to copy them in the clipboard. When you cut a block, all links connected to it are deleted as well.

    - Copy (Ctrl+C): Copy is used to place a copy of the selected item in the clipboard.

    - Paste (Ctrl+V): Paste places the content of the clipboard in the current diagram.

    - Delete (Delete): To delete blocks or links, select objects to be deleted and then the Delete menu item. When you delete a block, all links connected to it are deleted as well.

    - Select all (Ctrl+A): This menu item selects all the blocks in the current diagram.

    - Invert selection: This menu item inverts the current selection.

    - Block Parameters (Ctrl+B): This item opens the block configuration dialog for the current selected block. The configuration depends on the used block (see the Block Help menu item to obtain more information on its configuration).

    - Selection to superblock: This menu item converts a selection of blocks into a superblock.

179

- View

  - `Zoom in` (Ctrl+Plus): When you select this menu item, the diagram is zoomed in by a factor of 10.

  - `Zoom out` (Ctrl+Minus): When you select this menu item, the diagram is zoomed out by a factor of 10.

  - `Fit diagram to view`: When you select this menu item, the diagram is fit to the size of the current window.

  - `Normal 100%`: This menu item resize the diagram components at their normal displaying dimensions.

  - `Palette browser`: This menu item opens the palette browser.

  - `Diagram browser`: This menu item displays a window that lists the global properties of a diagram and all its objects (blocks and links).

  - `Viewport`: This menu item displays the Viewport window. With Viewport, you can move the working area onto a part of the diagram. You can zoom and unzoom parts of a diagram.

  - `Details`: This menu item displays a window by a selected block that lists properties of a block.

- Simulation:

  - `Setup`: In the main XCOS window, clicking the `Setup` menu item invokes the dialog box, which allows changing governing parameters.

  - `Execution Trace and Debug`: This menu item sets XCOS in debug mode. It opens a dialog box in which you can choose the debugging mode.

- Set Context: With this menu item, you obtain a dialog box where you can enter Scilab instructions for defining the symbolic XCOS parameters used in block definitions. These instructions will be evaluated each time the diagram is loaded. If you change the value of a symbolic XCOS parameter in the context, all the blocks that contains this symbolic parameter are updated when you click OK.

- Compile: This menu item compiles the block diagram. This doesn't need to be used because compilation is done automatically. Normally, a new compilation is not needed if only system parameters and internal states are modified.

- Modelica initialize: This menu item opens a specific dialog to the Modelica compiler, where you can see the components of a drawn model and also select the solvers.

- Start: This menu item starts the simulation. If the system has already been simulated, a dialog box appears where you can choose to continue, restart, or end the simulation.

- Stop: You can interrupt the simulation by clicking the Stop menu item. Block parameters can be changed and then simulation can be continued again.

- Format:

  - Rotate (Ctrl+R): Rotate allows you to turn a block on the left with an angle of 90°. Rotation affects all the selected blocks.

- Flip (Ctrl+F): To reverse the positions of the activation inputs and outputs set at the top and the bottom of a block, select a block and select the Flip menu item. This neither affects the order nor the position of the input and output event ports, which are numbered from left to right. Flipping affects all the selected blocks.

- Mirror (Ctrl+M): To reverse the positions of the (regular) inputs and outputs set on the left and the right of a block, select a block and select the Mirror menu item. This does not affect the order or the position of the input and output ports, which are numbered from top to bottom. Mirroring affects all the selected blocks.

- Show/Hide shadow: This menu item allows you to select 3D shapes for selected blocks and associated parameters.

- Align Blocks: When you select several blocks, it is possible to align them on a horizontal axis (top, bottom, and middle) or on vertical axis (left, right, center).

- Border Color: This menu item allows you to change the border color of the selected blocks.

- Fill Color: This menu item allows you to change the fill color of the selected blocks.

- Auto-Position: This menu item allows you to change the position of the block. First, select the block(s) and select the appropriate menu item or use the shortcut (P).

- Link Style: This menu item allows you to change the style of the link. First, select the link and select the appropriate menu item or use the shortcuts (H), (S), (V), and (O).

- Diagram background: This menu item allows you to change the background color.

- Grid: This menu item allows you to activate/de-activate the grid. With the grid, the block and link placement on the working area is easier and you obtain a more readable diagram.

- Help

  - Xcos Help (F1): This menu item opens the main help browser (if it is not open) on the XCOS help chapter.

  - Block Help: To get help on a XCOS block, select the block and then click this menu item.

  - XCOS Demonstrations: This item allows you to open some examples of XCOS diagrams.

  - About XCOS: The About XCOS item displays the current version of XCOS in a dialog.

8.7 Reconstructing the Bouncing Balls Example from Scratch

As an example, let's try to reconstruct the bouncing ball example discussed earlier from scratch. We will start with the option Demonstration blocks in the left pane of the XCOS palette window. Clicking this option presents three blocks, as shown in Figure 8-9. This example presents an XCOS simulation for visualizing three colored balls that bounce from the walls of their environment. One of the blocks, Bounce, produces the input, and another block, BounceXY, produces the output. For this reason, the former has two output ports on its right-hand side, whereas the latter has two input ports on its left-hand side.

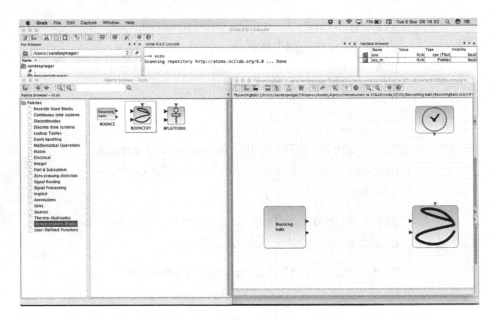

Figure 8-9. *Screenshot of XCOS's screen for the bouncing balls example*

Now, let's drag and drop the blocks titled Bounce and BounceXY on the XCOS editor window. The block Bounce has two output options, while BouncyXY has two input options. The BounceXY block actually has a third port for timer. The timer block can be collected from the section Event Handling. When all three are dragged and dropped at the editor window, you get a screen similar to the one shown in Figure 8-9.

As soon as you click one of the outputs of the Bounce block, a small, green colored square is highlighted around it. A connector can be clicked and dragged until the output of BouncyXY block and in this way a connection can be made. You also need a timing block, which controls the timing parameters for this simulation. This can be found in the Event Handing section of the left pane of the palette window.

Now these three blocks can be connected to obtain a connected diagram, as shown in Figure 8-10.

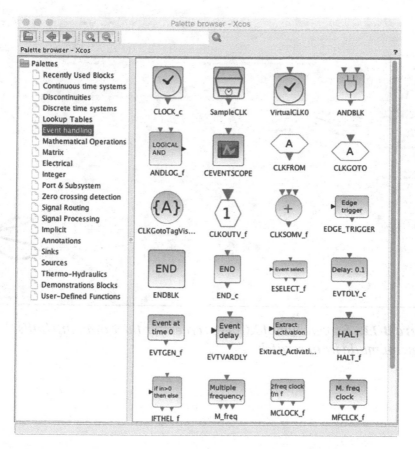

Figure 8-10. *XCOS palette window's section named "Event Handling"*

You can save this simulation with a name, say BouncingBalls.zcos
file. This simulation can then be made to run by clicking the Start button.
As opposed to the simulation model in Section 8.8, here you observe only
two balls bouncing off very fast.

The properties of a block can be changed by right-clicking a block
and choosing Block Parameters. For example, for all three blcoks in the
current simulation, you can see that a list of parameters (as shown in
Figures 8-12, 8-13, and 8-14) can be changed for running the simulation.

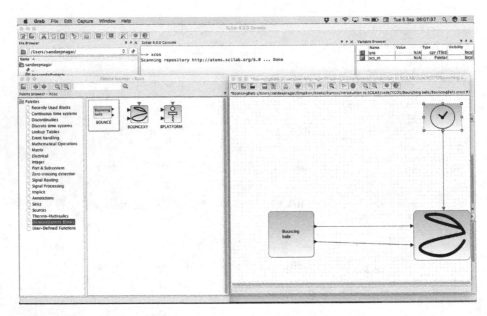

Figure 8-11. *Screenshot of XCOS's screen for the bouncing balls
diagram made from scratch*

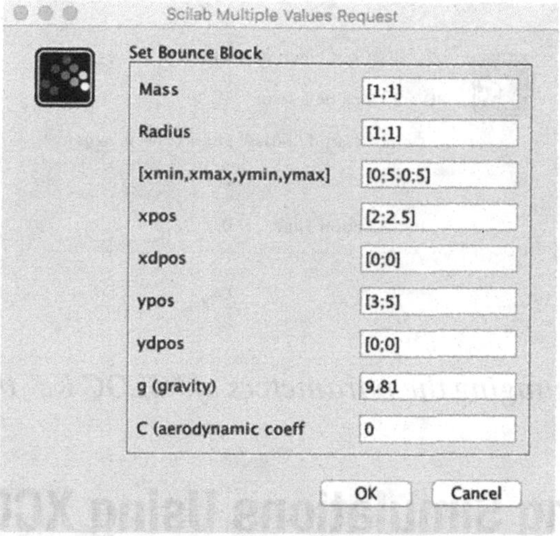

Figure 8-12. *Changing the parameters of the "bouncing balls" block*

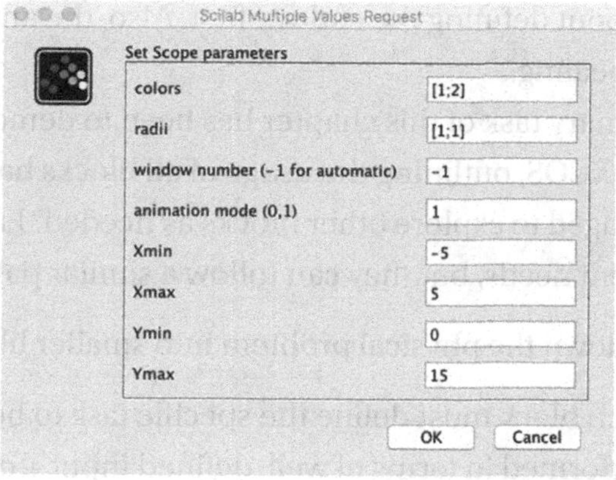

Figure 8-13. *Changing the parameters of "BOUNCEXY" block*

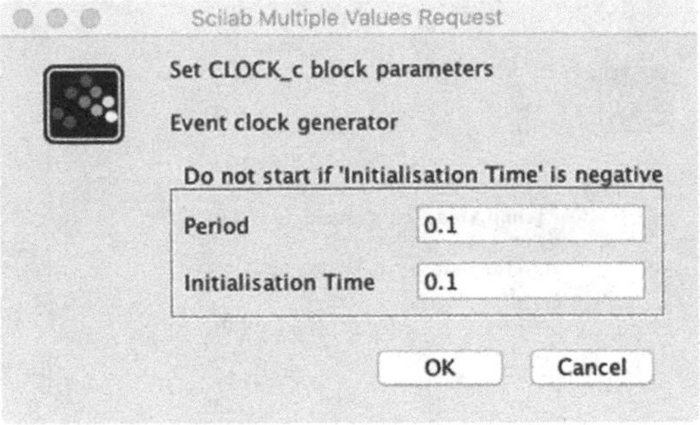

Figure 8-14. *Changing the parameters of "CLOC Kc" block*

8.8 Making Simulations Using XCOS

Simulation can prove to be a very powerful tool to both teach and quickly make scientific simulations. Since they are graphical in nature, the problems are quite intuitive to design. While teaching, you can avoid technical parts about defining the code at first. Also, the animations are quite visually appealing.

Since the primary task of this chapter has been to demonstrate a very primary usage of XCOS, outlining the usage of all blocks has been skipped. Users are encouraged to explore other blocks as needed. Each simulation model has different needs, but they can follow a similar pattern as follows:

- Break down the physical problem into smaller blocks

  - Each block must define the specific task to be performed in terms of well-defined input + opera-tion + output.

- Define the input and output for each block

  - Define the nature of each block in terms of number of channels, type of data at each channel, fre-quency of data, and so on

- Choose XCOS blocks that match these requirements and connect them as per model requirements

- Tune the model by changing the properties of the block(s) as required

Very complicated models would require defining a *superblock*. A superblock is a fully functional XCOS diagram that can be further made as a single block and used in other simulation diagrams. This way, you can break a bigger problem into blocks and their sub-blocks. This modular approach enables easy debugging provided all blocks are well labeled and th edeveloper is well informed about the capabilities of a particular block.

8.9 Summary

In this chapter, we have discussed the usage of the XCOS module available with Scilab. It provides the capabilities to program a physical problem in a graphical way. The physical problems must be chopped up in pieces of code. These blocks must explicitly define the input and output ports and the purpose as a code development process provided the developers know which block to find, where to find it, and how to use it. The documentation for each block is quite extensive and can be found using the help window. With this final chapter about XCOS, our introduction to Scilab is complete.

Index

Get the eBook for only $5!

Why limit yourself?

With most of our titles available in both PDF and ePUB format, you can access your content wherever and however you wish—on your PC, phone, tablet, or reader.

Since you've purchased this print book, we are happy to offer you the eBook for just $5.

To learn more, go to http://www.apress.com/companion or contact support@apress.com.

Apress®

Printed in the United States
By Bookmasters